Pioneers of Science

Oliver Lodge

Writat

This edition published in 2020

ISBN : 9789390439164 (Paperback)

Design and Setting By
Writat
(An Imprint of Vij Publishiing Group)
www.vijbooks.com
contact@vijpublishing.com

Table of Contents

PREFACE

THIS book takes its origin in a course of lectures on the history and progress of Astronomy arranged for me in the year 1887 by three of my colleagues (A.C.B., J.M., G.H.R.), one of whom gave the course its name.

The lectures having been found interesting, it was natural to write them out in full and publish.

If I may claim for them any merit, I should say it consists in their simple statement and explanation of scientific facts and laws. The biographical details are compiled from all readily available sources, there is no novelty or originality about them; though it is hoped that there may be some vividness. I have simply tried to present a living figure of each Pioneer in turn, and to trace his influence on the progress of thought.

I am indebted to many biographers and writers, among others to Mr. E.J.C. Morton, whose excellent set of lives published by the S.P.C.K. saved me much trouble in the early part of the course.

As we approach recent times the subject grows more complex, and the men more nearly contemporaries; hence the biographical aspect diminishes and the scientific treatment becomes fuller, but in no case has it been allowed to become technical and generally unreadable.

To the friends (C.C.C., F.W.H.M., E.F.R.) who with great kindness have revised the proofs, and have indicated places where the facts could be made more readily intelligible by a clearer statement, I express my genuine gratitude.

UNIVERSITY COLLEGE, LIVERPOOL,
November, 1892.

PART I

FROM DUSK TO DAYLIGHT

DATES AND SUMMARY OF FACTS FOR LECTURE I

Physical Science of the Ancients. Thales 640 B.C., Anaximander 610 B.C., PYTHAGORAS 600 B.C., Anaxagoras 500 B.C., Eudoxus 400 B.C., ARISTOTLE 384 B.C., Aristarchus 300 B.C., ARCHIMEDES 287 B.C., Eratosthenes 276 B.C., HIPPARCHUS 160 B.C., Ptolemy 100 A.D.

Science of the Middle Ages. Cultivated only among the Arabs; largely in the forms of astrology, alchemy, and algebra.

Return of Science to Europe. Roger Bacon 1240, Leonardo da Vinci 1480, (Printing 1455), Columbus 1492, Copernicus 1543.

A sketch of Copernik's life and work. Born 1473 at Thorn in Poland. Studied mathematics at Bologna. Became an ecclesiastic. Lived at Frauenburg near mouth of Vistula. Substituted for the apparent motion of the heavens the real motion of the earth. Published tables of planetary motions. Motion still supposed to be in epicycles. Worked out his ideas for 36 years, and finally dedicated his work to the Pope. Died just as his book was printed, aged 72, a century before the birth of Newton. A colossal statue by Thorwaldsen erected at Warsaw in 1830.

LECTURE I

COPERNICUS AND THE MOTION OF THE EARTH

THE ordinary run of men live among phenomena of which they know nothing and care less. They see bodies fall to the earth, they hear sounds, they kindle fires, they see the heavens roll above them, but of the causes and inner working of the whole they are ignorant, and with their ignorance they are content.

"Understand the structure of a soap-bubble?" said a cultivated literary man whom I know; "I wouldn't cross the street to know it!"

And if this is a prevalent attitude now, what must have been the attitude in ancient times, when mankind was emerging from savagery, and when history seems composed of harassments by wars abroad and revolutions at home? In the most violently disturbed times indeed, those with which ordinary history is mainly occupied, science is quite impossible. It needs as its condition, in order to flourish, a fairly quiet, untroubled state, or else a cloister or university removed from the din and bustle of the political and commercial world. In such places it has taken its rise, and in such peaceful places and quiet times true science will continue to be cultivated.

The great bulk of mankind must always remain, I suppose, more or less careless of scientific research and scientific result, except in so far as it affects their modes of locomotion, their health and pleasure, or their purse.

But among a people hurried and busy and preoccupied, some in the pursuit of riches, some in the pursuit of pleasure, and some, the majority, in the struggle for existence, there arise in every generation, here and there, one or two great souls—men who seem of another age and country, who look upon the bustle and feverish activity and are not infected by it, who watch others achieving prizes of riches and pleasure and are not disturbed, who look on the world and the universe they are born in with quite other eyes. To them it appears not as a bazaar to buy and to sell in; not as a ladder to scramble up (or down) helter-skelter without knowing whither or why; but as a fact—a great and mysterious fact—to be pondered over, studied, and perchance in some small measure understood. By the multitude these men were sneered at as eccentric or feared as supernatural. Their calm, clear, contemplative attitude seemed either insane or diabolic; and accordingly they have been pitied as enthusiasts or killed as blasphemers. One of these great souls may have been a prophet or preacher, and have called to his generation to bethink them of why and what they were, to struggle less and meditate more, to search for things of true value and not for dross. Another has been a poet or musician, and has uttered in words or in song thoughts dimly possible to many men, but by them unutterable and left inarticulate. Another has been influenced still more *directly* by the universe around him, has felt at times overpowered by the mystery and solemnity of it all, and has been impelled by a force stronger than himself to study it, patiently, slowly, diligently; content if he could gather a few crumbs of the great harvest of knowledge, happy if he could grasp some great generalization or wide-embracing law, and so in some small measure enter into the mind and thought of the Designer of all this wondrous frame of things.

These last have been the men of science, the great and heaven-born men of science; and they are few. In our own day, amid the throng of inventions, there are a multitude of small men using the name of science but working for their own ends, jostling and scrambling just as they would jostle and scramble in any other trade or profession. These may be workers, they may and do advance knowledge, but they are never pioneers. Not to them is it given to open out great tracts of unexplored territory, or to view the promised land as from a mountain-top. Of them we shall not speak; we will concern ourselves only with the greatest, the epoch-making men, to whose life and work we and all who come after them owe so much. Such a man was Thales. Such was Archimedes, Hipparchus, Copernicus. Such pre-eminently was Newton.

Now I am not going to attempt a history of science. Such a work in ten lectures would be absurd. I intend to pick out a few salient names here and there, and to study these in some detail, rather than by attempting to deal with too many to lose individuality and distinctness.

We know so little of the great names of antiquity, that they are for this purpose scarcely suitable. In some departments the science of the Greeks was remarkable, though it is completely overshadowed by their philosophy; yet it was largely based on what has proved to be a wrong method of procedure, viz the introspective and conjectural, rather than the inductive and experimental methods. They investigated Nature by studying their own minds, by considering the meanings of words, rather than by studying things and recording phenomena. This wrong (though by no means, on the face of it, absurd) method was not pursued exclusively, else would their science have been valueless, but the influence it had was such as materially to detract from the value of their speculations and discoveries. For when truth and falsehood are inextricably woven into a statement, the truth is as hopelessly hidden as if it had never been stated, for we have no criterion to distinguish the false from the true.

FIG. 1.—Archimedes.

Besides this, however, many of their discoveries were ultimately lost to the world, some, as at Alexandria, by fire—the bigoted work of a Mohammedan conqueror—some by irruption of barbarians; and all were buried so long and so completely by the night of the dark ages, that they had to be rediscovered almost as absolutely and completely as though they had never been. Some of the names of antiquity we shall have occasion to refer to; so I have arranged some of them in chronological order on page 4, and as a representative one I may specially emphasize Archimedes, one of the greatest men of science there has ever been, and the father of physics.

The only effective link between the old and the new science is afforded by the Arabs. The dark ages come as an utter gap in the scientific history of Europe, and for more than a thousand years there was not a scientific man of note except in Arabia; and with the Arabs knowledge was so mixed up with magic and enchantment that one cannot contemplate it with any degree of satisfaction, and little real progress was made. In some of the *Waverley Novels* you can realize the state of matters in these times; and you know how the only approach to science is through some Arab sorcerer or astrologer, maintained usually by a monarch, and consulted upon all great occasions, as the oracles were of old.

In the thirteenth century, however, a really great scientific man appeared, who may be said to herald the dawn of modern science in Europe. This man was Roger Bacon. He cannot be said to do more than herald it, however, for we must wait two hundred years for the next name of great magnitude; moreover he was isolated, and so far in advance of his time that he left no followers. His own work suffered from the prevailing ignorance, for he was persecuted and imprisoned, not for the commonplace and natural reason that he frightened the Church, but merely because he was eccentric in his habits and knew too much.

The man I spoke of as coming two hundred years later is Leonardo da Vinci. True he is best known as an artist, but if you read his works you will come to the conclusion that he was the most scientific artist who ever lived. He teaches the laws of perspective (then new), of light and shade, of colour, of the equilibrium of bodies, and of a multitude of other matters where science touches on art—not always quite correctly according to modern ideas, but in beautiful and precise language. For clear and conscious power, for wide-embracing knowledge and skill, Leonardo is one of the most remarkable men that ever lived.

About this time the tremendous invention of printing was achieved, and Columbus unwittingly discovered the New World. The middle of the next century must be taken as the real dawn of modern science; for the year 1543 marks the publication of the life-work of Copernicus.

FIG. 2.—Leonardo da Vinci.

Nicolas Copernik was his proper name. Copernicus is merely the Latinized form of it, according to the then prevailing fashion. He was born at Thorn, in Polish Prussia, in 1473. His father is believed to have been a German. He graduated at Cracow as doctor in arts and medicine, and was destined for the ecclesiastical profession. The details of his life are few; it seems to have been quiet and uneventful, and we know very little about it. He was instructed in astronomy at Cracow, and learnt mathematics at Bologna. Thence he went to Rome, where he was made Professor of Mathematics; and soon afterwards he went into orders. On his return home, he took charge of the principal church in his native place, and became a canon. At Frauenburg, near the mouth of the Vistula, he lived the remainder of his life. We find him reporting on coinage for the Government, but otherwise he does not appear as having entered into the life of the times.

He was a quiet, scholarly monk of studious habits, and with a reputation which drew to him several earnest students, who received *vivâ voce* instruction from him; so, in study and meditation, his life passed.

He compiled tables of the planetary motions which were far more correct than any which had hitherto appeared, and which remained serviceable for long afterwards. The Ptolemaic system of the heavens, which had been the orthodox system all through the Christian era, he endeavoured to improve and simplify by the hypothesis that the sun was the centre of the system instead of the earth; and the first consequences of this change he worked out for many years, producing in the end a great book: his one life-work. This famous work, "De Revolutionibus Orbium Cœlestium," embodied all his painstaking calculations, applied his new system to each of the bodies in the solar system in succession, and treated besides of much other recondite matter. Towards the close of his life it was put into type. He can scarcely be said to have lived to see it appear, for he was stricken with paralysis before its completion; but a printed copy was brought to his bedside and put into his hands, so that he might just feel it before he died.

FIG. 3.—Copernicus.

That Copernicus was a giant in intellect or power—such as had lived in the past, and were destined to live in the near future—I see no reason whatever to believe. He was just a quiet, earnest, patient, and God-fearing man, a deep student, an unbiassed thinker, although with no specially brilliant or striking gifts; yet to him it was given to effect such a revolution in the whole course of man's thoughts as is difficult to parallel.

You know what the outcome of his work was. It proved—he did not merely speculate, he proved—that the earth is a planet like the others, and that it revolves round the sun.

Yes, it can be summed up in a sentence, but what a revelation it contains. If you have never made an effort to grasp the full significance of this discovery you will not appreciate it. The doctrine is very familiar to us now, we have heard it, I suppose, since we were four years old, but can you realize it? I know it was a long time before I could. Think of the solid earth, with trees and houses, cities and countries, mountains and seas—think of the vast tracts of land in Asia, Africa, and America—and then picture the whole mass spinning like a top, and rushing along its annual course round the sun at the rate of nineteen miles every second.

Were we not accustomed to it, the idea would be staggering. No wonder it was received with incredulity. But the difficulties of the conception are not only physical, they are still more felt from the speculative and theological points of view. With this last, indeed, the reconcilement cannot be considered complete even yet. Theologians do not, indeed, now *deny* the fact of the earth's subordination in the scheme of the universe, but many of them ignore it and pass it by. So soon as the Church awoke to a perception of the tremendous and revolutionary import of the new doctrines, it was bound to resist them or be false to its traditions. For the whole tenor of men's thought must have been changed had they accepted it. If the earth were not the central and all-important body in the universe, if the sun and planets and stars were not attendant and subsidiary lights, but were other worlds larger and perhaps superior to ours, where was man's place in the universe? and where were the doctrines they had maintained as irrefragable? I by no means assert that the new doctrines were really utterly irreconcilable with the more essential parts of the old dogmas, if only theologians had had patience and genius enough to consider the matter calmly. I suppose that in that case they might have reached the amount of reconciliation at present attained, and not only have left scientific truth in peace to spread as it could, but might perhaps themselves have joined the band of earnest students and workers, as so many of the higher Catholic clergy do at the present day.

But this was too much to expect. Such a revelation was not to be accepted in a day or in a century—the easiest plan was to treat it as a heresy, and try to crush it out.

Not in Copernik's life, however, did they perceive the dangerous tendency of the doctrine—partly because it was buried in a ponderous and learned treatise not likely to be easily understood; partly, perhaps, because its propounder was himself an ecclesiastic; mainly because he was a patient and judicious man, not given to loud or intolerant assertion, but content to state his views in quiet conversation, and to let them gently spread for thirty years before he published them. And, when he did publish them, he used the happy device of dedicating his great book to the Pope, and a cardinal bore the expense of printing it. Thus did the Roman Church stand sponsor to a system of truth against which it was destined in the next century to hurl its anathemas, and to inflict on its conspicuous adherents torture, imprisonment, and death.

To realize the change of thought, the utterly new view of the universe, which the Copernican theory introduced, we must go back to preceding ages, and try to recall the views which had been held as probable concerning the form of the earth and the motion of the heavenly bodies.

FIG. 4.—Homeric Cosmogony.

The earliest recorded notion of the earth is the very natural one that it is a flat area floating in an illimitable ocean. The sun was a god who drove his chariot across the heavens once a day; and Anaxagoras was threatened with death and punished with banishment for teaching that the sun was only a ball of fire, and that it might perhaps be as big as the country of Greece. The obvious difficulty as to how the sun got back to the east again every morning was got over—not by the conjecture that he went back in the dark, nor by the idea that there was a fresh sun every day; though, indeed, it was once believed that the moon was created once a month, and periodically cut up into stars—but by the doctrine that in the northern part of the earth was a high range of mountains, and that the sun travelled round on the surface of the sea behind these. Sometimes, indeed, you find a representation of the sun being rowed round in a boat. Later on it was perceived to be necessary that the sun should be able to travel beneath the earth, and so the earth was supposed to be supported on pillars or on roots, or to be a dome-shaped body floating in air—much like Dean Swift's island of Laputa. The elephant and tortoise of the Hindu earth are, no doubt, emblematic or typical, not literal.

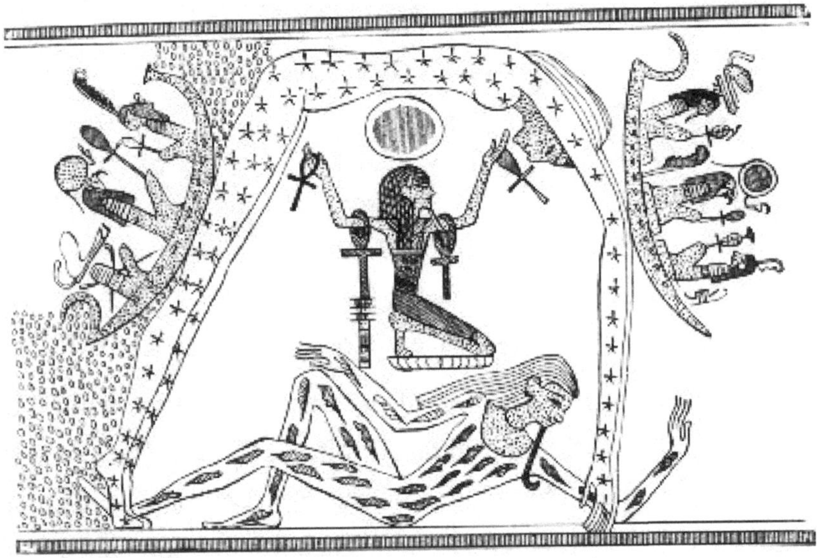

FIG. 5.—Egyptian Symbol of the Universe.

The earth a figure with leaves, the heaven a figure with stars, the principle of equilibrium and support, the boats of the rising and setting sun.

Aristotle, however, taught that the earth must be a sphere, and used all the orthodox arguments of the present children's geography-books about the way you see ships at sea, and about lunar eclipses.

To imagine a possible antipodes must, however, have been a tremendous difficulty in the way of this conception of a sphere, and I scarcely suppose that any one can at that time have contemplated the possibility of such upside-down regions being inhabited. I find that intelligent children invariably feel the greatest difficulty in realizing the existence of inhabitants on the opposite side of the earth. Stupid children, like stupid persons in general, will of course believe anything they are told, and much good may the belief do them; but the kind of difficulties felt by intelligent and thoughtful children are most instructive, since it is quite certain that the early philosophers must have encountered and overcome those very same difficulties by their own genius.

FIG. 6.—Hindoo Earth.

However, somehow or other the conception of a spherical earth was gradually grasped, and the heavenly bodies were perceived all to revolve round it: some moving regularly, as the stars, all fixed together into one spherical shell or firmament; some moving irregularly and apparently anomalously—these irregular bodies were therefore called planets [or wanderers]. Seven of them were known, viz. Moon, Mercury, Venus, Sun, Mars, Jupiter, Saturn, and there is little doubt that this number seven, so suggested, is the origin of the seven days of the week.

The above order of the ancient planets is that of their supposed distance from the earth. Not always, however, are they thus quoted by the ancients: sometimes the sun is supposed nearer than Mercury or Venus. It has always been known that the moon was the nearest of the heavenly bodies; and some rough notion of its distance was current. Mars, Jupiter, and Saturn were placed in that order because that is the order of their apparent motions, and it was natural to suppose that the slowest moving bodies were the furthest off.

The order of the days of the week shows what astrologers considered to be the order of the planets; on their system of each successive hour of the day being ruled over by the successive planets taken in order. The diagram (fig. 7) shows that if the Sun rule the first hour of a certain day (thereby giving its name to the day) Venus will rule the second hour, Mercury the third, and so on; the Sun will thus be found to rule the eighth, fifteenth, and twenty-second hour of that day, Venus the twenty-third, and Mercury the twenty-fourth hour; so the Moon will rule the first hour of the next day, which will therefore be Monday. On the same principle (numbering round the hours successively, with the arrows) the first hour of the next day will be found to be ruled by Mars, or by the Saxon deity corresponding thereto; the first hour of the day after, by Mercury (*Mercredi*), and so on (following the straight lines of the pattern).

The order of the planets round the circle counter-clockwise, *i.e.* the direction of their proper motions, is that quoted above in the text.

To explain the motion of the planets and reduce them to any sort of law was a work of tremendous difficulty. The greatest astronomer of ancient times was Hipparchus, and to him the system known as the Ptolemaic system is no doubt largely due. But it was delivered to the world mainly by Ptolemy, and goes by his name. This was a fine piece of work, and a great advance on anything that had gone before; for although it is of course saturated with error, still it is based on a large substratum of truth. Its superiority to all the previously mentioned systems is obvious. And it really did in its more developed form describe the observed motions of the planets.

Each planet was, in the early stages of this system, as taught, say, by Eudoxus, supposed to be set in a crystal sphere, which revolved so as to carry the planet with it. The sphere had to be of crystal to account for the visibility of other planets and the stars through it. Outside the seven planetary spheres, arranged one inside the other, was a still larger one in which were set the stars. This was believed to turn all the others, and was called the *primum mobile*. The whole system was supposed to produce, in its revolution, for the few privileged to hear the music of the spheres, a sound as of some magnificent harmony.

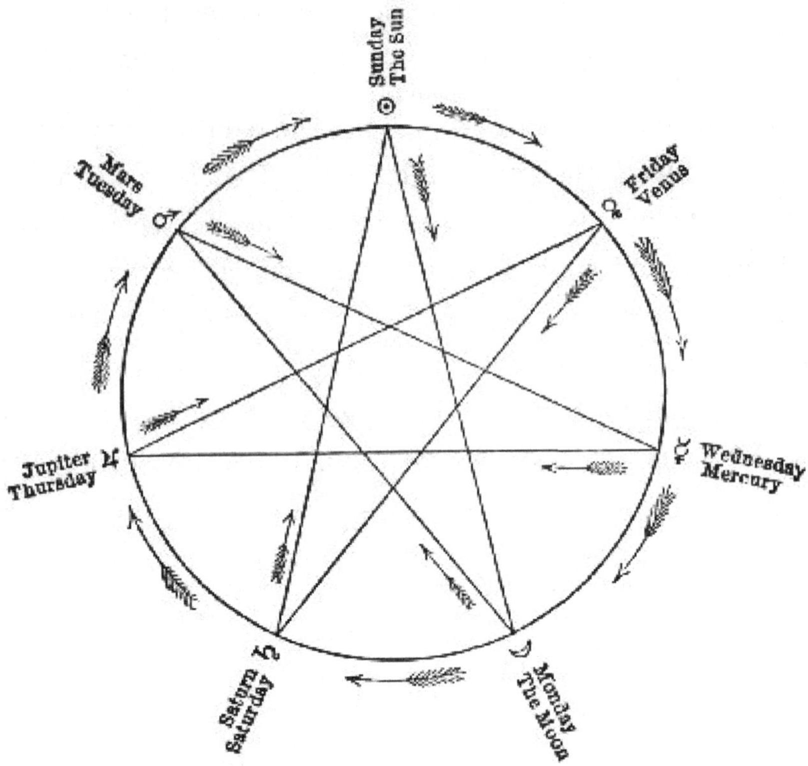

FIG. 7.—Order of ancient planets corresponding to the days of the week.

The enthusiastic disciples of Pythagoras believed that their master was privileged to hear this noble chant; and far be it from us to doubt that the rapt and absorbing pleasure of contemplating the harmony of nature, to a man so eminently great as Pythagoras, must be truly and adequately represented by some such poetic conception.

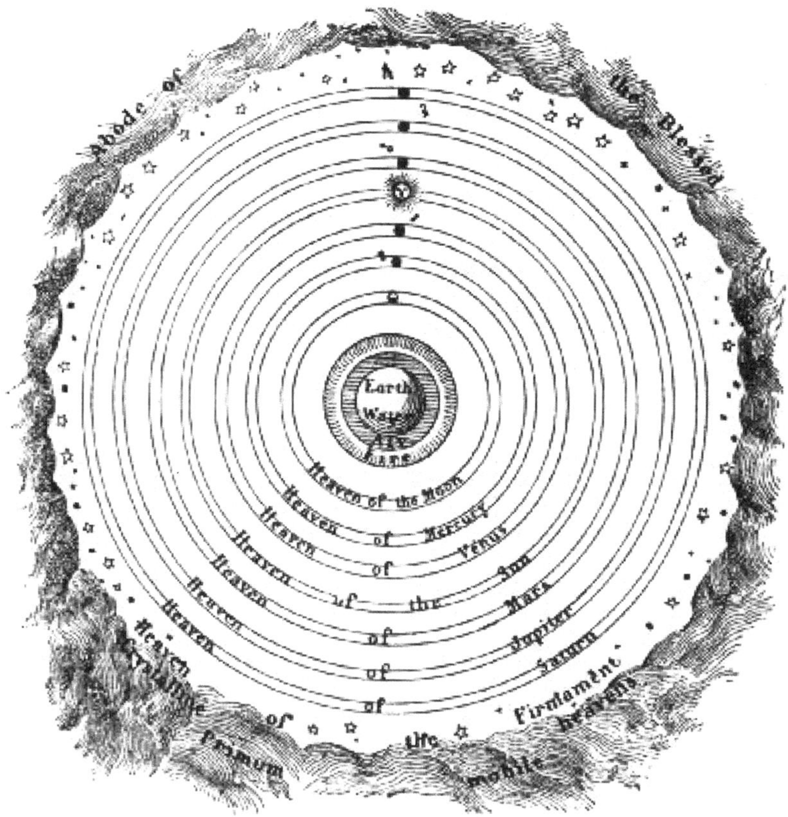

FIG. 8.—Ptolemaic system.

The precise kind of motion supposed to be communicated from the *primum mobile* to the other spheres so as to produce the observed motions of the planets was modified and improved by various philosophers until it developed into the epicyclic train of Hipparchus and of Ptolemy.

It is very instructive to observe a planet (say Mars or Jupiter) night after night and plot down its place with reference to the fixed stars on a celestial globe or star-map. Or, instead of direct observation by alignment with known stars, it is easier to look out its right ascension and declination in *Whitaker's Almanac*, and plot those down. If this be done for a year or two, it will be found that the motion of the planet is by no means regular, but that though on the whole it advances it sometimes is stationary and sometimes goes back.[1]

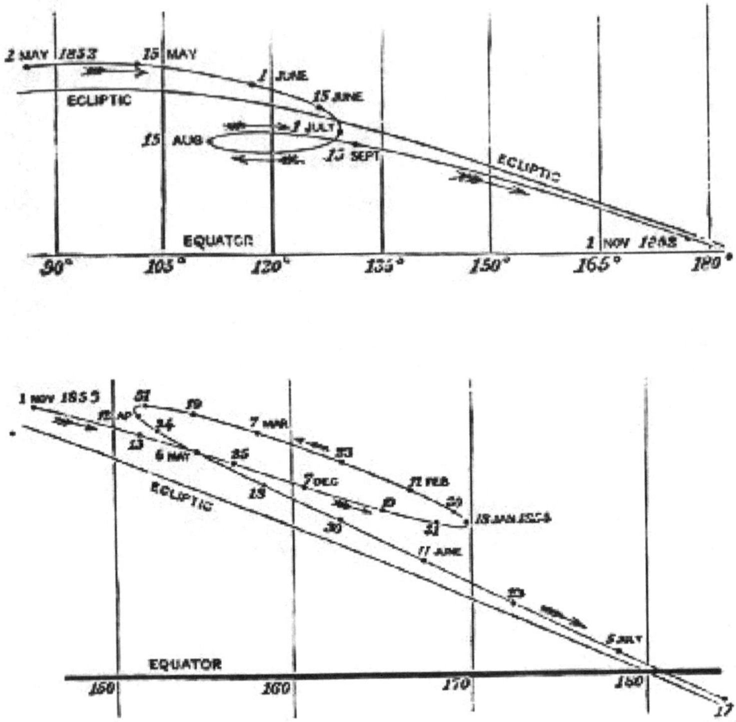

FIG. 9.—Specimens of Apparent paths of Venus and of Mars among the stars.

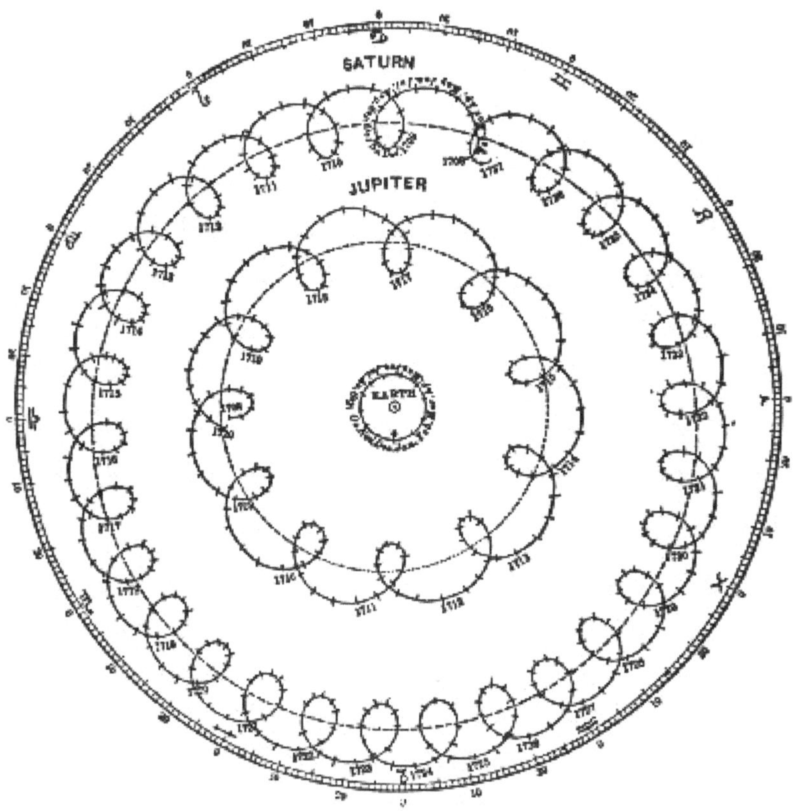

FIG. 10.—Apparent epicyclic orbits of Jupiter and Saturn; the Earth being supposed fixed at the centre, with the Sun revolving in a small circle. A loop is made by each planet every year.

These "stations" and "retrogressions" of the planets were well known to the ancients. It was not to be supposed for a moment that the crystal spheres were subject to any irregularity, neither was uniform circular motion to be readily abandoned; so it was surmised that the main sphere carried, not the planet itself, but the centre or axis of a subordinate sphere, and that the planet was carried by this. The minor sphere could be allowed to revolve at a different uniform pace from the main sphere, and so a curve of some complexity could be obtained.

A curve described in space by a point of a circle or sphere, which itself is carried along at the same time, is some kind of cycloid; if the centre of the tracing circle travels along a straight line, we get the ordinary cycloid, the curve traced in air by a nail on a coach-wheel; but if the centre of the tracing circle be carried round another circle the curve described is called an epicycloid. By such curves the planetary stations and retrogressions could be explained. A large sphere would have to revolve once for a "year" of the particular planet, carrying with it a subsidiary sphere in which the planet was fixed; this latter sphere revolving once for a "year" of the earth. The actual looped curve thus described is depicted for Jupiter and Saturn in the annexed diagram (fig. 10.)

It was long ago perceived that real material spheres were unnecessary; such spheres indeed, though possibly transparent to light, would be impermeable to comets: any other epicyclic gearing would serve, and as a mere description of the motion it is simpler to think of a system of jointed bars, one long arm carrying a shorter arm, the two revolving at different rates, and the end of the short one carrying the planet. This does all that is needful for the first approximation to a planet's motion. In so far as the motion cannot be thus truly stated, the short arm may be supposed to carry another, and that another, and so on, so that the resultant motion of the planet is compounded of a large number of circular motions of different periods; by this device any required amount of complexity could be attained. We shall return to this at greater length in Lecture III.

The main features of the motion, as shown in the diagram, required only two arms for their expression; one arm revolving with the average motion of the planet, and the other revolving with the apparent motion of the sun, and always pointing in the same direction as the single arm supposed to carry the sun. This last fact is of course because the motion to be represented does not really belong to the planet at all, but to the earth, and so all the main epicyclic motions for the superior planets were the same. As for the inferior planets (Mercury and Venus) they only appear to oscillate like the bob of a pendulum about the sun, and so it is very obvious that they must be really revolving round it. An ancient Egyptian system perceived this truth; but the Ptolemaic system imagined them to revolve round the earth like the rest, with an artificial system of epicycles to prevent their ever getting far away from the neighbourhood of the sun.

It is easy now to see how the Copernican system explains the main features of planetary motion, the stations and retrogressions, quite naturally and without any complexity.

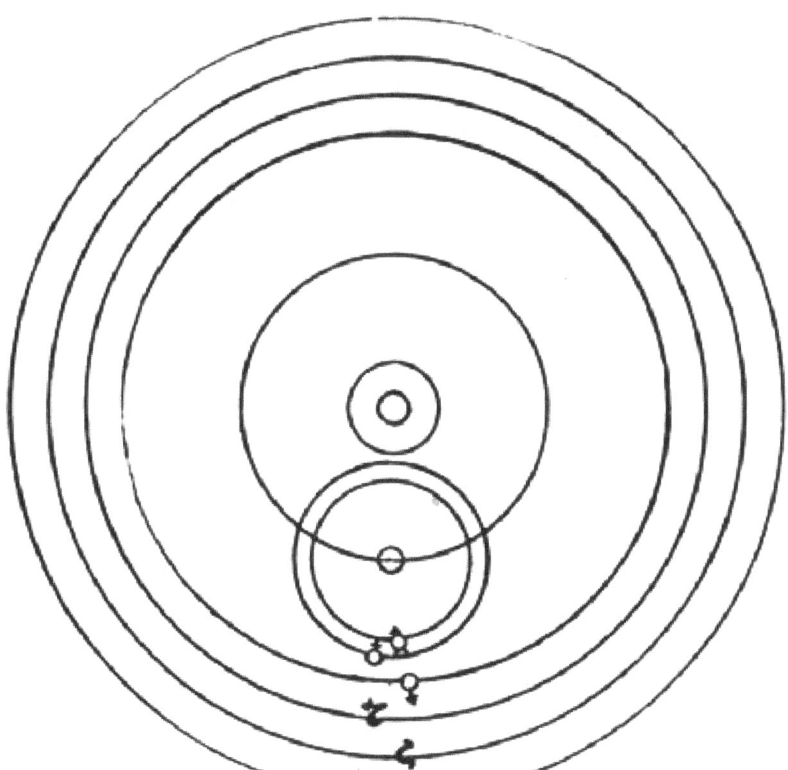

FIG. 11.—Egyptian system.

Let the outer circle represent the orbit of Jupiter, and the inner circle the orbit of the earth, which is moving faster than Jupiter (since Jupiter takes 4332 days to make one revolution); then remember that the apparent position of Jupiter is referred to the infinitely distant fixed stars and refer to fig. 12.

Let E_1, E_2, &c., be successive positions of the earth; J_1, J_2, &c., corresponding positions of Jupiter. Produce the lines $E_1 J_1$, $E_2 J_2$, &c., to an enormously greater circle outside, and it will be seen that the termination of these lines, representing apparent positions of Jupiter among the stars, advances while the earth goes from E_1 to E_3; is almost stationary from somewhere about E_3 to E_4; and recedes from E_4 to E_5; so that evidently the recessions of Jupiter are only apparent, and are due to the orbital motion of the earth. The apparent complications in the path of Jupiter, shown in Fig. 10, are seen to be caused simply by the motion of the earth, and to be thus completely and easily explained.

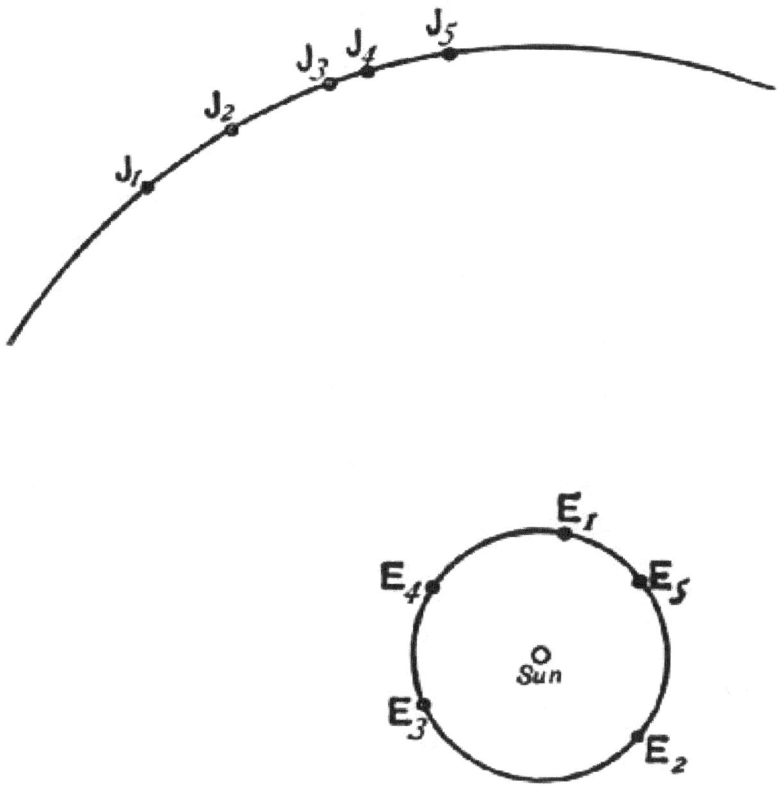

FIG. 12.—True orbits of Earth and Jupiter.

The same thing for an inferior planet, say Mercury, is even still more easily seen (*vide* figure 13).

The motion of Mercury is direct from M'' to M''', retrograde from M''' to M'', and stationary at M'' and M'''. It appears to oscillate, taking 72·5 days for its direct swing, and 43·5 for its return swing.

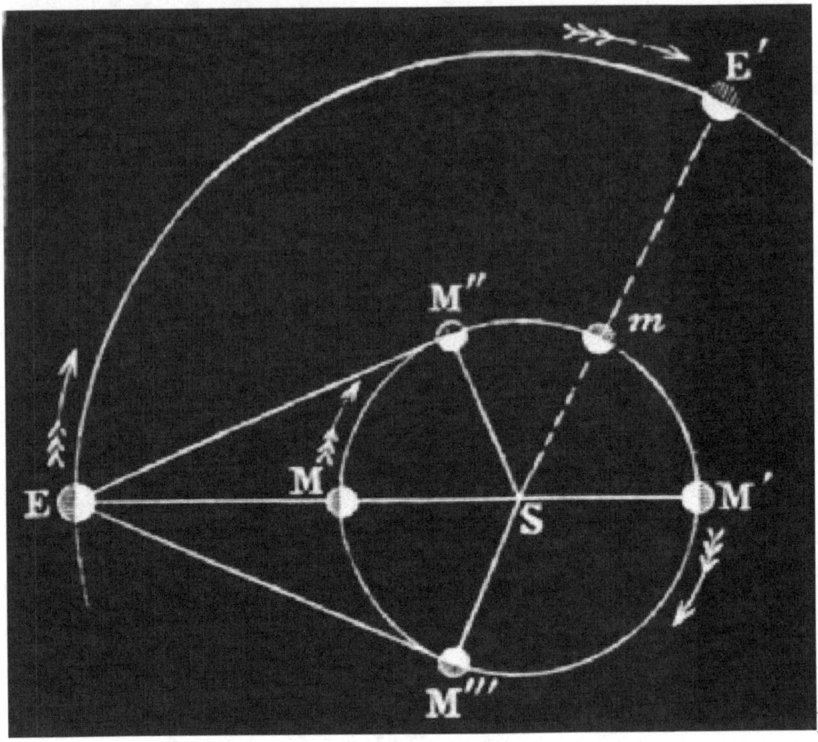

FIG. 13.—Orbit of Mercury and Earth.

On this system no artificiality is required to prevent Mercury's ever getting far from the sun: the radius of its orbit limits its real and apparent excursions. Even if the earth were stationary, the motions of Mercury and Venus would not be *essentially* modified, but the stations and retrogressions of the superior planets, Mars, Jupiter, &c., would wholly cease.

The complexity of the old mode of regarding apparent motion may be illustrated by the case of a traveller in a railway train unaware of his own motion. It is as though trees, hedges, distant objects, were all flying past him and contorting themselves as you may see the furrows of a ploughed field do when travelling, while you yourself seem stationary amidst it all. How great a simplicity would be introduced by the hypothesis that, after all, these things might be stationary and one's self moving.

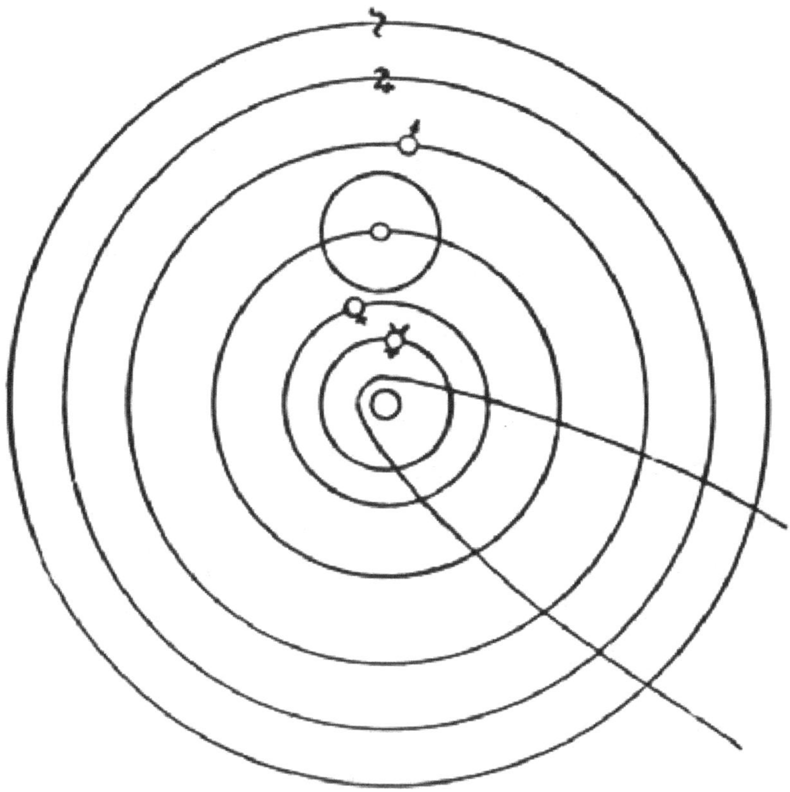

FIG. 14.—Copernican system as frequently represented. But the cometary orbit is a much later addition, and no attempt is made to show the relative distances of the planets.

Now you are not to suppose that the system of Copernicus swept away the entire doctrine of epicycles; that doctrine can hardly be said to be swept away even now. As a description of a planet's motion it is not incorrect, though it is geometrically cumbrous. If you describe the motion of a railway train by stating that every point on the rim of each wheel describes a cycloid with reference to the earth, and a circle with reference to the train, and that the motion of the train is compounded of these cycloidal and circular motions, you will not be saying what is false, only what is cumbrous.

The Ptolemaic system demanded large epicycles, depending on the motion of the earth, these are what Copernicus overthrew; but to express the minuter details of the motion smaller epicycles remained, and grew more and more complex as observations increased in accuracy, until a greater man than either Copernicus or Ptolemy, viz. Kepler, replaced them all by a simple ellipse.

One point I must not omit from this brief notice of the work of Copernicus. Hipparchus had, by most sagacious interpretation of certain observations of his, discovered a remarkable phenomenon called the precession of the equinoxes. It was a discovery of the first magnitude, and such as would raise to great fame the man who should have made it in any period of the world's history, even the present. It is scarcely expressible in popular language, and without some technical terms; but I can try.

The plane of the earth's orbit produced into the sky gives the apparent path of the sun throughout a year. This path is known as the ecliptic, because eclipses only happen when the moon is in it. The sun keeps to it accurately, but the planets wander somewhat above and below it (fig. 9), and the moon wanders a good deal. It is manifest, however, in order that there may be an eclipse of any kind, that a straight line must be able to be drawn through earth and moon and sun (not necessarily through their centres of course), and this is impossible unless some parts of the three bodies are in one plane, viz. the ecliptic, or something very near it. The ecliptic is a great circle of the sphere, and is usually drawn on both celestial and terrestrial globes.

The earth's equator also produced into the sky, where it may still be called the equator (sometimes it is awkwardly called "the equinoctial"), gives another great circle inclined to the ecliptic and cutting it at two opposite points, labelled respectively ♈ and ♎, and together called "the equinoxes." The reason for the name is that when the sun is in that part of the ecliptic it is temporarily also on the equator, and hence is symmetrically situated with respect to the earth's axis of rotation, and consequently day and night are equal all over the earth.

Well, Hipparchus found, by plotting the position of the sun for a long time,[2] that these points of intersection, or equinoxes, were not stationary from century to century, but slowly moved among the stars, moving as it were to meet the sun, so that he gets back to one of these points again 20 minutes 23¼ seconds before it has really completed a revolution, *i.e.* before the true year is fairly over. This slow movement forward of the goal-post is called precession—the precession of the equinoxes. (One result of it is to shorten our years by about 20 minutes each; for the shortened period has to be called a year, because it is on the position of the sun with respect to the earth's axis that our seasons depend.) Copernicus perceived that, assuming the motion of the earth, a clearer account of this motion could be given. The ordinary approximate statement concerning the earth's axis is that it remains parallel to itself, *i.e.* has a fixed direction as the earth moves round the sun. But if, instead of being thus fixed, it be supposed to have a slow movement of revolution, so that it traces out a cone in the course of about 26,000 years, then, since the equator of course goes with it, the motion of its intersection with the fixed ecliptic is so far accounted for. That is to say, the precession of the equinoxes is seen to be dependent on, and caused by, a slow conical movement of the earth's axis.

The prolongation of each end of the earth's axis into the sky, or the celestial north and south poles, will thus slowly trace out an approximate circle among the stars; and the course of the north pole during historic time is exhibited in the annexed diagram.

It is now situated near one of the stars of the Lesser Bear, which we therefore call the Pole star; but not always was it so, nor will it be so in the future. The position of the north pole 4000 years ago is shown in the figure; and a revolution will be completed in something like 26,000 years.[3]

FIG. 15.—Slow movement of the north pole in a circle among the stars. (Copied from Sir R. Ball.)

This perception of the conical motion of the earth's axis was a beautiful generalization of Copernik's, whereby a multitude of facts were grouped into a single phenomenon. Of course he did not explain the motion of the axis itself. He stated the fact that it so moved, and I do not suppose it ever struck him to seek for an explanation.

An explanation was given later, and that a most complete one; but the idea even of seeking for it is a brilliant and striking one: the achievement of the explanation by a single individual in the way it actually was accomplished is one of the most astounding things in the history of science; and were it not that the same individual accomplished a dozen other things, equally and some still more extraordinary, we should rank that man as one of the greatest astronomers that ever lived.

As it is, he is Sir Isaac Newton.

We are to remember, then, as the life-work of Copernicus, that he placed the sun in its true place as the centre of the solar system, instead of the earth; that he greatly simplified the theory of planetary motion by this step, and also by the simpler epicyclic chain which now sufficed, and which he worked out mathematically; that he exhibited the precession of the equinoxes (discovered by Hipparchus) as due to a conical motion of the earth's axis; and that, by means of his simpler theory and more exact planetary tables, he reduced to some sort of order the confused chaos of the Ptolemaic system, whose accumulation of complexity and of outstanding errors threatened to render astronomy impossible by the mere burden of its detail.

There are many imperfections in his system, it is true; but his great merit is that he dared to look at the facts of Nature with his own eyes, unhampered by the prejudice of centuries. A system venerable with age, and supported by great names, was universally believed, and had been believed for centuries. To doubt this system, and to seek after another and better one, at a time when all men's minds were governed by tradition and authority, and when to doubt was sin—this required a great mind and a high character. Such a mind and such a character had this monk of Frauenburg. And it is interesting to notice that the so-called religious scruples of smaller and less truly religious men did not affect Copernicus; it was no dread of consequences to one form of truth that led him to delay the publication of the other form of truth specially revealed to him. In his dedication he says:—

"If there be some babblers who, though ignorant of all mathematics, take upon them to judge of these things, and dare to blame and cavil at my work, because of some passage of Scripture which they have wrested to their own purpose, I regard them not, and will not scruple to hold their judgment in contempt."

I will conclude with the words of one of his biographers (Mr. E.J.C. Morton):—

"Copernicus cannot be said to have flooded with light the dark places of nature—in the way that one stupendous mind subsequently did—but still, as we look back through the long vista of the history of science, the dim Titanic figure of the old monk seems to rear itself out of the dull flats around it, pierces with its head the mists that overshadow them, and catches the first gleam of the rising sun,

> "'... like some iron peak, by the Creator
> Fired with the red glow of the rushing
> morn.'"

DATES AND SUMMARY OF FACTS FOR LECTURE II

Copernicus lived from 1473 to 1543, and was contemporary with Paracelsus and Raphael.

Tycho Brahé	from 1546 to 1601.
Kepler	from 1571 to 1630.
Galileo	from 1564 to 1642.
Gilbert	from 1540 to 1603.
Francis Bacon	from 1561 to 1626.
Descartes	from 1596 to 1650.

A sketch of Tycho Brahé's life and work. Tycho was a Danish noble, born on his ancestral estate at Knudstorp, near Helsinborg, in 1546. Adopted by his uncle, and sent to the University of Copenhagen to study law. Attracted to astronomy by the occurrence of an eclipse on its predicted day, August 21st, 1560. Began to construct astronomical instruments, especially a quadrant and a sextant. Observed at Augsburg and Wittenberg. Studied alchemy, but was recalled to astronomy by the appearance of a new star. Overcame his aristocratic prejudices, and delivered a course of lectures at Copenhagen, at the request of the king. After this he married a peasant girl. Again travelled and observed in Germany. In 1576 was sent for to Denmark by Frederick II., and established in the island of Huen, with an endowment enabling him to devote his life to astronomy. Built Uraniburg, furnished it with splendid instruments, and became the founder of accurate instrumental astronomy. His theories were poor, but his observations were admirable. In 1592 Frederick died, and five years later, Tycho was impoverished and practically banished. After wandering till 1599, he was invited to Prague by the Emperor Rudolf, and there received John Kepler among other pupils. But the sentence of exile was too severe, and he died in 1601, aged 54 years.

A man of strong character, untiring energy, and devotion to accuracy, his influence on astronomy has been immense.

LECTURE II

TYCHO BRAHÉ AND THE EARLIEST OBSERVATORY

WE have seen how Copernicus placed the earth in its true position in the solar system, making it merely one of a number of other worlds revolving about a central luminary. And observe that there are two phenomena to be thus accounted for and explained: first, the diurnal revolution of the heavens; second, the annual motion of the sun among the stars.

The effect of the diurnal motion is conspicuous to every one, and explains the rising, southing, and setting of the whole visible firmament. The effect of the annual motion, *i.e.* of the apparent annual motion, of the sun among the stars, is less obvious, but it may be followed easily enough by observing the stars visible at any given time of evening at different seasons of the year. At midnight, for instance, the position of the sun is definite, viz. due north always, but the constellation which at that time is due south or is rising or setting varies with the time of year; an interval of one month producing just the same effect on the appearance of the constellations as an interval of two hours does (because the day contains twice as many hours as the year contains months), *e.g.* the sky looks the same at midnight on the 1st of October as it does at 10 p.m. on the 1st of November.

All these simple consequences of the geocentric as opposed to the heliocentric point of view were pointed out by Copernicus, in addition to his greater work of constructing improved planetary tables on the basis of his theory. But it must be admitted that he himself felt the hypothesis of the motion of the earth to be a difficulty. Its acceptance is by no means such an easy and childish matter as we are apt now to regard it, and the hostility to it is not at all surprising. The human race, after having ridiculed and resisted the truth for a long time, is apt to end in accepting it so blindly and unimaginatively as to fail to recognize the real achievement of its first propounders, or the difficulties which they had to overcome. The majority of men at the present day have grown accustomed to hear the motion of the earth spoken of: their acceptance of it means nothing: the attitude of the paradoxer who denies it is more intelligent.

It is not to be supposed that the idea of thus explaining some of the phenomena of the heavens, especially the daily motion of the entire firmament, by a diurnal rotation of the earth had not struck any one. It was often at this time referred to as the Pythagorean theory, and it had been taught, I believe, by Aristarchus. But it was new to the modern world, and it had the great weight of Aristotle against it. Consequently, for long after Copernicus, only a few leading spirits could be found to support it, and the long-established venerable Ptolemaic system continued to be taught in all Universities.

The main objections to the motion of the earth were such as the following:—

1. The motion is unfelt and difficult to imagine.

That it is unfelt is due to its uniformity, and can be explained mechanically. That it is difficult to imagine is and remains true, but a most important lesson we have to learn is that difficulty of conception is no valid argument against reality.

2. That the stars do not alter their relative positions according to the season of the year, but the constellations preserve always the same aspect precisely, even to careful measurement.

This is indeed a difficulty, and a great one. In June the earth is 184 million miles away from where it was in December: how can we see precisely the same fixed stars? It is not possible, unless they are at a practically infinite distance. That is the only answer that can be given. It was the tentative answer given by Copernicus. It is the correct answer. Not only from every position of the earth, but from every planet of the solar system, the same constellations are visible, and the stars have the same aspect. The whole immensity of the solar system shrinks to practically a point when confronted with the distance of the stars.

Not, however, so entirely a speck as to resist the terrific accuracy of the present century, and their microscopic relative displacement with the season of the year has now at length been detected, and the distance of many thereby measured.

3. That, if the earth revolved round the sun, Mercury and Venus ought to show phases like the moon.

So they ought. Any globe must show phases if it live nearer the sun than we do and if we go round it, for we shall see varying amounts of its illuminated half. The only answer that Copernicus could give to this was that they might be difficult to see without extra powers of sight, but he ventured to predict that the phases would be seen if ever our powers of vision should be enhanced.

4. That if the earth moved, or even revolved on its own axis, a stone or other dropped body ought to be left far behind.

This difficulty is not a real one, like the two last, and it is based on an ignorance of the laws of mechanics, which had not at that time been formulated. We know now that a ball dropped from a high tower, so far from lagging, drops a minute trifle *in front* of the foot of a perpendicular, because the top of the tower is moving a trace faster than the bottom, by reason of the diurnal rotation. But, ignoring this, a stone dropped from the lamp of a railway carriage drops in the centre of the floor, whether the carriage be moving steadily or standing still; a slant direction of fall could only be detected if the carriage were being accelerated or if the brake were applied. A body dropped from a moving carriage shares the motion of the carriage, and starts with that as its initial velocity. A ball dropped from a moving balloon does not simply drop, but starts off in whatever direction the car was moving, its motion being immediately modified by gravity, precisely in the same way as that of a thrown ball is modified. This is, indeed, the whole philosophy of throwing—to drop a ball from a moving carriage. The carriage is the hand, and, to throw far, a run is taken and the body is jerked forward; the arm is also moved as rapidly as possible on the shoulder as pivot. The fore-arm can be moved still faster, and the wrist-joint gives yet another motion: the art of throwing is to bring all these to bear at the same instant, and then just as they have all attained their maximum velocity to let the ball go. It starts off with the initial velocity thus imparted, and is abandoned to gravity. If the vehicle were able to continue its motion steadily, as a balloon does, the ball when let go from it would appear to the occupant simply to drop; and it would strike the ground at a spot vertically under the moving vehicle, though by no means vertically below the place where it started. The resistance of the air makes observations of this kind inaccurate, except when performed inside a carriage so that the air shares in the motion. Otherwise a person could toss and catch a ball out of a train window just as well as if he were stationary; though to a spectator outside he would seem to be using great skill to throw the ball in the parabola adapted to bring it back to his hand.

The same circumstance enhances the apparent difficulty of the circus rider's jumping feats. All he has to do is to jump up and down on the horse; the forward motion which carries him through hoops belongs to him by virtue of the motion of the horse, without effort on his part.

Thus, then, it happens that a stone dropped sixteen feet on the earth appears to fall straight down, although its real path in space is a very flat trajectory of nineteen miles base and sixteen feet height; nineteen miles being the distance traversed by the earth every second in the course of its annual journey round the sun.

No wonder that it was thought that bodies must be left behind if the earth was subject to such terrific speed as this. All that Copernicus could suggest on this head was that perhaps the atmosphere might help to carry things forward, and enable them to keep pace with the earth.

There were thus several outstanding physical difficulties in the way of the acceptance of the Copernican theory, besides the Biblical difficulty.

It was quite natural that the idea of the earth's motion should be repugnant, and take a long time to sink into the minds of men; and as scientific progress was vastly slower then than it is now, we find not only all priests but even some astronomers one hundred years afterwards still imagining the earth to be at rest. And among them was a very eminent one, Tycho Brahé.

It is interesting to note, moreover, that the argument about the motion of the earth being contrary to Scripture appealed not only to ecclesiastics in those days, but to scientific men also; and Tycho Brahé, being a man of great piety, and highly superstitious also, was so much influenced by it, that he endeavoured to devise some scheme by which the chief practical advantages of the Copernican system could be retained, and yet the earth be kept still at the centre of the whole. This was done by making all the celestial sphere, with stars and everything, rotate round the earth once a day, as in the Ptolemaic scheme; and then besides this making all the planets revolve round the sun, and this to revolve round the earth. Such is the Tychonic system.

So far as *relative* motion is concerned it comes to the same thing; just as when you drop a book you may say either that the earth rises to meet the book, or that the book falls to meet the earth. Or when a fly buzzes round your head, you may say that you are revolving round the fly. But the absurdity of making the whole gigantic system of sun and planets and stars revolve round our insignificant earth was too great to be swallowed by other astronomers after they had once had a taste of the Copernican theory; and accordingly the Tychonic system died a speedy and an easy death at the same time as its inventor.

Wherein then lay the magnitude of the man?—not in his theories, which were puerile, but in his observations, which were magnificent. He was the first observational astronomer, the founder of the splendid system of practical astronomy which has culminated in the present Greenwich Observatory.

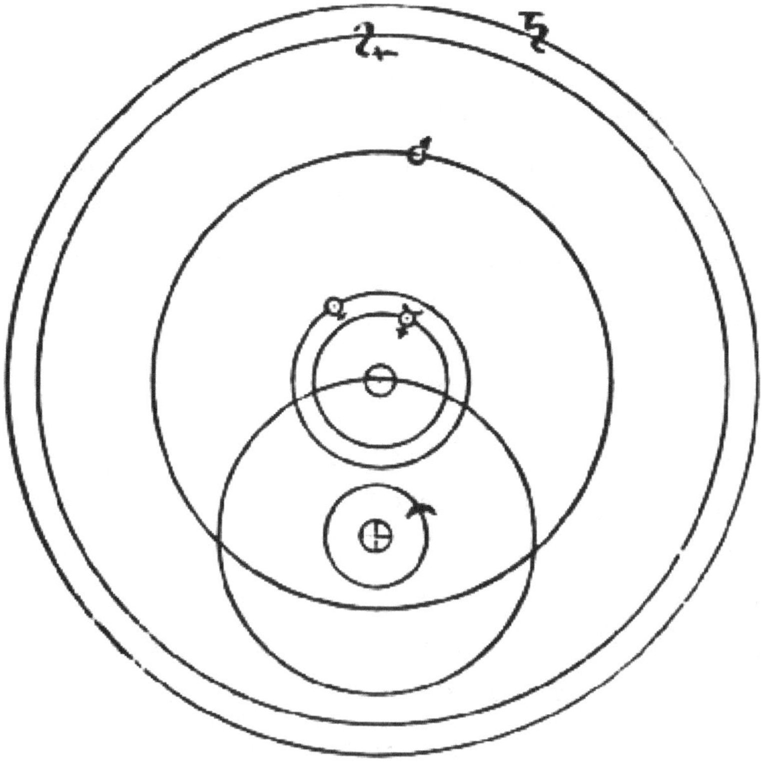

FIG. 16.—Tychonic system showing the sun with all the planets revolving round the earth.

Up to Tycho the only astronomical measurements had been of the rudest kind. Copernicus even improved upon what had gone before, with measuring rules made with his own hands. Ptolemy's observations could never be trusted to half a degree. Tycho introduced accuracy before undreamed of, and though his measurements, reckoned by modern ideas, are of course almost ludicrously rough (remember no such thing as a telescope or microscope was then dreamed of), yet, estimated by the era in which they were made, they are marvels of accuracy, and not a single mistake due to carelessness has ever been detected in them. In fact they may be depended on almost to minutes of arc, *i.e.* to sixtieths of a degree.

For certain purposes connected with the proper motion of stars they are still appealed to, and they served as the certain and trustworthy data for succeeding generations of theorists to work upon. It was long, indeed, after Tycho's death before observations approaching in accuracy to his were again made.

In every sense, therefore, he was a pioneer: let us proceed to trace his history.

Born the eldest son of a noble family—"as noble and ignorant as sixteen undisputed quarterings could make them," as one of his biographers says—in a period when, even more than at present, killing and hunting were the only natural aristocratic pursuits, when all study was regarded as something only fit for monks, and when science was looked at askance as something unsavoury, useless, and semi-diabolic, there was little in his introduction to the world urging him in the direction where his genius lay. Of course he was destined for a soldier; but fortunately his uncle, George Brahé, a more educated man than his father, having no son of his own, was anxious to adopt him, and though not permitted to do so for a time, succeeded in getting his way on the birth of a second son, Steno—who, by the way, ultimately became Privy Councillor to the King of Denmark.

Tycho's uncle gave him what he would never have got at home—a good education; and ultimately put him to study law. At the age of thirteen he entered the University of Copenhagen, and while there occurred the determining influence of his life.

An eclipse of the sun in those days was not regarded with the cold-blooded inquisitiveness or matter-of-fact apathy, according as there is or is not anything to be learnt from it, with which such an event is now regarded. Every occurrence in the heavens was then believed to carry with it the destiny of nations and the fate of individuals, and accordingly was of surpassing interest. Ever since the time of Hipparchus it had been possible for some capable man here and there to predict the occurrence of eclipses pretty closely. The thing is not difficult. The prediction was not, indeed, to the minute and second, as it is now; but the day could usually be hit upon pretty accurately some time ahead, much as we now manage to hit upon the return of a comet—barring accidents; and the hour could be predicted as the event approached.

Well, the boy Tycho, among others, watched for this eclipse on August 21st, 1560; and when it appeared at its appointed time, every instinct for the marvellous, dormant in his strong nature, awoke to strenuous life, and he determined to understand for himself a science permitting such wonderful possibilities of prediction. He was sent to Leipzig with a tutor to go on with his study of law, but he seems to have done as little law as possible: he spent all his money on books and instruments, and sat up half the night studying and watching the stars.

In 1563 he observed a conjunction of Jupiter and Saturn, the precursor, and *cause* as he thought it, of the great plague. He found that the old planetary tables were as much as a month in error in fixing this event, and even the Copernican tables were several days out; so he formed the resolve to devote his life to improving astronomical tables. This resolve he executed with a vengeance. His first instrument was a jointed ruler with sights for fixing the position of planets with respect to the stars, and observing their stations and retrogressions. By thus measuring the angles between a planet and two fixed stars, its position can be plotted down on a celestial map or globe.

FIG. 17.—Portrait of Tycho.

In 1565 his uncle George died, and made Tycho his heir. He returned to Denmark, but met with nothing but ridicule and contempt for his absurd drivelling away of time over useless pursuits. So he went back to Germany—first to Wittenberg, thence, driven by the plague, to Rostock.

Here his fiery nature led him into an absurd though somewhat dangerous adventure. A quarrel at some feast, on a mathematical point, with a countryman, Manderupius, led to the fixing of a duel, and it was fought with swords at 7 p.m. at the end of December, when, if there was any light at all, it must have been of a flickering and unsatisfactory nature. The result of this insane performance was that Tycho got his nose cut clean off.

He managed however to construct an artificial one, some say of gold and silver, some say of putty and brass; but whatever it was made of there is no doubt that he wore it for the rest of his life, and it is a most famous feature. It excited generally far more interest than his astronomical researches. It is said, moreover, to have very fairly resembled the original, but whether this remark was made by a friend or by an enemy I cannot say. One account says that he used to carry about with him a box of cement to apply whenever his nose came off, which it periodically did.

About this time he visited Augsburg, met with some kindred and enlightened spirits in that town, and with much enthusiasm and spirit constructed a great quadrant. These early instruments were tremendous affairs. A great number of workmen were employed upon this quadrant, and it took twenty men to carry it to its place and erect it. It stood in the open air for five years, and then was destroyed by a storm. With it he made many observations.

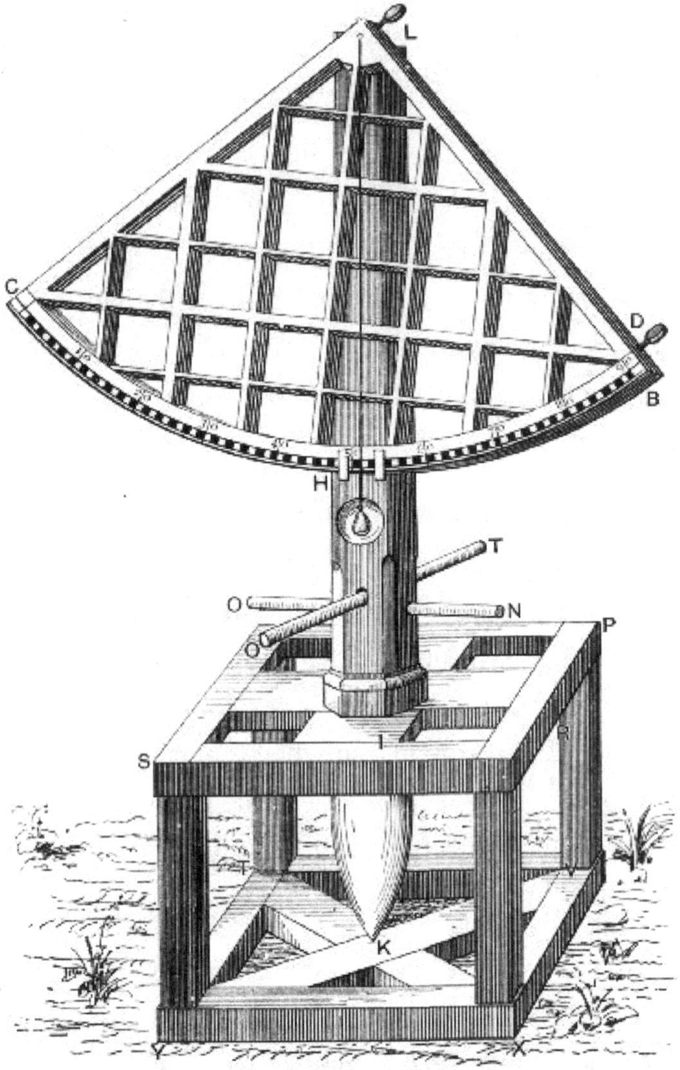

FIG. 18.—Early out-door quadrant of Tycho; for observing altitudes by help of the sights *D*, *L* and the plumb line.

On his return to Denmark in 1571, his fame preceded him, and he was much better received; and in order to increase his power of constructing instruments he took up the study of alchemy, and like the rest of the persuasion tried to make gold. The precious metals were by many old philosophers considered to be related in some way to the heavenly bodies: silver to the moon, for instance—as we still see by the name lunar caustic applied to nitrate of silver; gold to the sun, copper to Mars, lead to Saturn. Hence astronomy and alchemy often went together. Tycho all his life combined a little alchemy with his astronomical labours, and he constructed a wonderful patent medicine to cure all disorders, which had as wide a circulation in Europe in its time as Holloway's pills; he gives a tremendous receipt for it, with liquid gold and all manner of ingredients in it; among them, however, occurs a little antimony—a well-known sudorific—and to this, no doubt, whatever efficacy the medicine possessed was due.

So he might have gone on wasting his time, were it not that in November, 1572, a new star made its appearance, as they have done occasionally before and since. On the average one may say that about every fifty years a new star of fair magnitude makes its temporary appearance. They are now known to be the result of some catastrophe or collision, whereby immense masses of incandescent gas are produced. This one seen by Tycho became as bright as Jupiter, and then died away in about a year and a half. Tycho observed all its changes, and endeavoured to measure its distance from the earth, with the result that it was proved to belong to the region of the fixed stars, at an immeasurable distance, and was not some nearer and more trivial phenomenon.

He was asked by the University of Copenhagen to give a course of lectures on astronomy; but this was a step he felt some aristocratic aversion to, until a little friendly pressure was brought to bear upon him by a request from the king, and delivered they were.

He now seems to have finally thrown off his aristocratic prejudices, and to have indulged himself in treading on the corns of nearly all the high and mighty people he came into contact with. In short, he became what we might now call a violent Radical; but he was a good-hearted man, nevertheless, and many are the tales told of his visits to sick peasants, of his consulting the stars as to their fate—all in perfect good faith—and of the medicines which he concocted and prescribed for them.

The daughter of one of these peasants he married, and very happy the marriage seems to have been.

FIG. 19.—Map of Denmark, showing the island of Huen.
Walker & Boutallse.

Now comes the crowning episode in Tycho's life. Frederick II., realizing how eminent a man they had among them, and how much he could do if only he had the means—for we must understand that Tycho, though of good family and well off, was by no means what we would call a wealthy man—Frederick II. made him a splendid and enlightened offer. The offer was this: that if Tycho would agree to settle down and make his astronomical observations in Denmark, he should have an estate in Norway settled upon him, a pension of £400 a year for life, a site for a large observatory, and £20,000 to build it with.

FIG. 20.—Uraniburg.

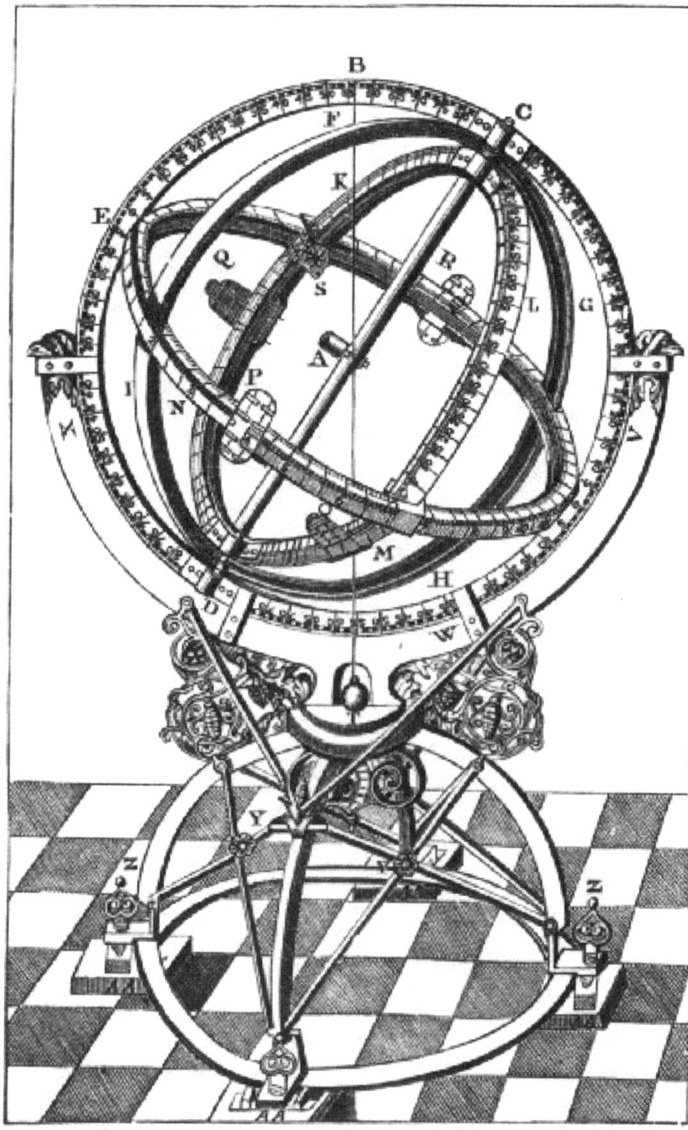

FIG. 21.—Astrolabe. An old instrument with sights for marking the positions of the celestial bodies roughly. A sort of skeleton celestial globe.

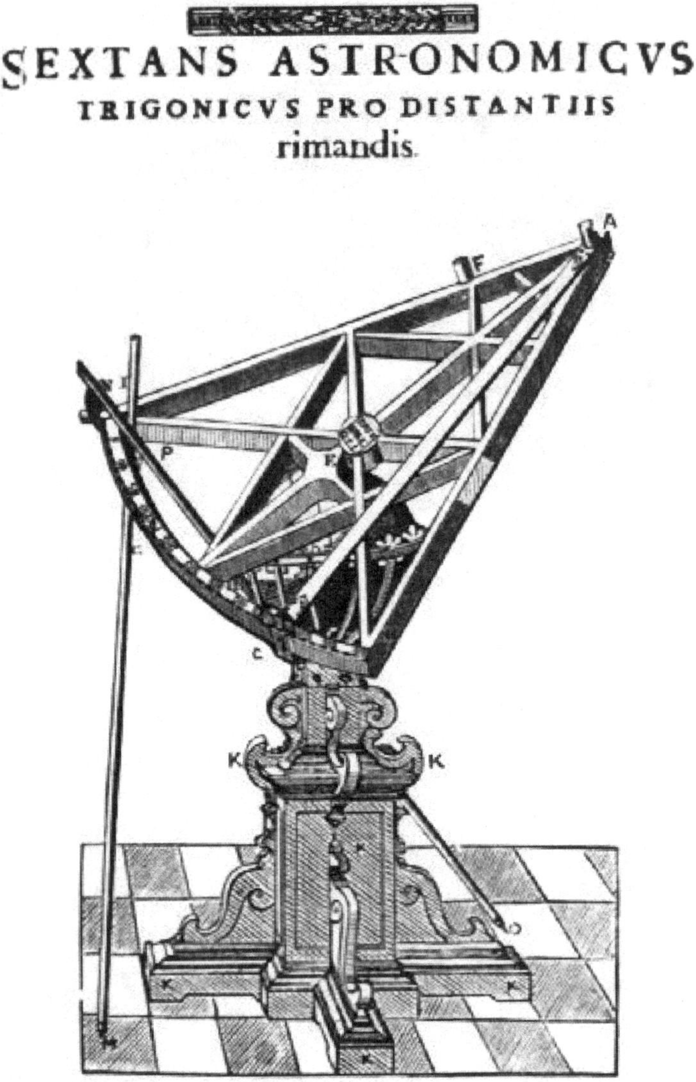

FIG. 22.—Tycho's large sextant; for measuring the angular distance between two bodies by direct sighting.

Well, if ever money was well spent, this was. By its means Denmark before long headed the nations of Europe in the matter of science—a thing it has not done before or since. The site granted was the island of Huen, between Copenhagen and Elsinore; and here the most magnificent observatory ever built was raised, and called Uraniburg—the castle of the heavens. It was built on a hill in the centre of the island, and included gardens, printing shops, laboratory, dwelling-houses, and four observatories—all furnished with the most splendid instruments that Tycho could devise, and that could then be constructed. It was decorated with pictures and sculptures of eminent men, and altogether was a most gorgeous place. £20,000 no doubt went far in those days, but the original grant was supplemented by Tycho himself, who is said to have spent another equal sum out of his own pocket on the place.

QVADRANS MAXIMVS CHALI
BEUS QUADRATO INCLUSUS, ET
HORIZONTI AZIMUTHALI CHALYBEO
INSISTENS.

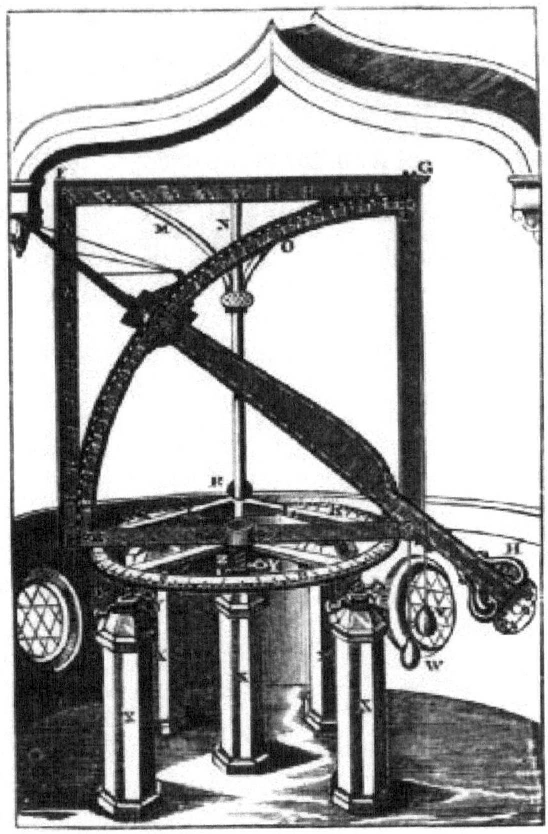

FIG. 23.—The Quadrant in Uraniburg; or altitude and azimuth instrument.

For twenty years this great temple of science was continually worked in by him, and he soon became the foremost scientific man in Europe. Philosophers, statesmen, and occasionally kings, came to visit the great astronomer, and to inspect his curiosities.

QVADRANS MVRALIS
SIVE TICHONICUS.

FIG. 24.—Tycho's form of transit circle.

The method of utilising the extremely uniform rotation of the earth by watching the planets and stars as they cross the meridian, and recording their times of transit; observing also at the same time their meridian altitudes (see observer *F*), was the invention of Tycho, and constitutes his greatest achievement. His method is followed to this day in all observatories.

FIG. 25.—A modern transit circle, showing essentially the same parts as in Tycho's instrument, viz. the observer watching the transit, the clock, the recorder of the observation, and the graduated circle; the latter to be read by a second observer.

And very wholesome for some of these great personages must have been the treatment they met with. For Tycho was no respecter of persons. His humbly-born wife sat at the head of the table, whoever was there; and he would snub and contradict a chancellor just as soon as he would a serf. Whatever form his pride may have taken when a youth, in his maturity it impelled him to ignore differences of rank not substantially justified, and he seemed to take a delight in exposing the ignorance of shallow titled persons, to whom contradiction and exposure were most unusual experiences.

For sick peasants he would take no end of trouble, and went about doctoring them for nothing, till he set all the professional doctors against him; so that when his day of misfortune came, as come it did, their influence was not wanting to help to ruin one who spoilt their practice, and whom they derided as a quack.

But some of the great ignorant folk who came to visit his temple of science, and to inspect its curiosities, felt themselves insulted—not always without reason. He kept a tame maniac in the house, named Lep, and he used to regard the sayings of this personage as oracular, presaging future events, and far better worth listening to than ordinary conversation. Consequently he used to have him at his banquets and feed him himself; and whenever Lep opened his mouth to speak, every one else was peremptorily ordered to hold his tongue, so that Lep's words might be written down. In fact it was something like an exaggerated edition of Betsy Trotwood and Mr. Dick.

"It must have been an odd dinner party" (says Prof. Stuart), "with this strange, wild, terribly clever man, with his red hair and brazen nose, sometimes flashing with wit and knowledge, sometimes making the whole company, princes and servants alike, hold their peace and listen humbly to the ravings of a poor imbecile."

To people he despised he did not show his serious instruments. He had other attractions, in the shape of a lot of toy machinery, little windmills, and queer doors, and golden globes, and all manner of ingenious tricks and automata, many of which he had made himself, and these he used to show them instead; and no doubt they were well enough pleased with them. Those of the visitors, however, who really cared to see and understand his instruments, went away enchanted with his genius and hospitality.

I may, perhaps, be producing an unfair impression of imperiousness and insolence. Tycho was fiery, no doubt, but I think we should wrong him if we considered him insolent. Most of the nobles of his day were haughty persons, accustomed to deal with serfs, and very likely to sneer at and trample on any meek man of science whom they could easily despise. So Tycho was not meek; he stood up for the honour of his science, and paid them back in their own coin, with perhaps a little interest. That this behaviour was not worldly-wise is true enough, but I know of no commandment enjoining us to be worldly-wise.

If we knew more about his so-called imbecile *protégé* we should probably find some reason for the interest which Tycho took in him. Whether he was what is now called a "clairvoyant" or not, Tycho evidently regarded his utterances as oracular, and of course when one is receiving what may be a revelation from heaven it is natural to suppress ordinary conversation.

Among the noble visitors whom he received and entertained, it is interesting to notice James I. of England, who spent eight days at Uraniburg on the occasion of his marriage with Anne of Denmark in 1590, and seems to have been deeply impressed by his visit.

Among other gifts, James presented Tycho with a dog (depicted in Fig. 24), and this same animal was subsequently the cause of trouble. For it seems that one day the Chancellor of Denmark, Walchendorf, brutally kicked the poor beast; and Tycho, who was very fond of animals, gave him a piece of his mind in no measured language. Walchendorf went home determined to ruin him. King Frederick, however, remained his true friend, doubtless partly influenced thereto by his Queen Sophia, an enlightened woman who paid many visits to Uraniburg, and knew Tycho well. But unfortunately Frederick died; and his son, a mere boy, came to the throne.

Now was the time for the people whom Tycho had offended, for those who were jealous of his great fame and importance, as well as for those who cast longing eyes on his estate and endowments. The boy-king, too, unfortunately paid a visit to Tycho, and, venturing upon a decided opinion on some recondite subject, received a quiet setting down which he ill relished.

Letters written by Tycho about this time are full of foreboding. He greatly dreads having to leave Uraniburg, with which his whole life has for twenty years been bound up. He tries to comfort himself with the thought that, wherever he is sent, he will have the same heavens and the same stars over his head.

Gradually his Norwegian estate and his pension were taken away, and in five years poverty compelled him to abandon his magnificent temple, and to take a small house in Copenhagen.

Not content with this, Walchendorf got a Royal Commission appointed to inquire into the value of his astronomical labours. This sapient body reported that his work was not only useless, but noxious; and soon after he was attacked by the populace in the public street.

Nothing was left for him now but to leave the country, and he went into Germany, leaving his wife and instruments to follow him whenever he could find a home for them.

His wanderings in this dark time—some two years—are not quite clear; but at last the enlightened Emperor of Bohemia, Rudolph II., invited him to settle in Prague. Thither he repaired, a castle was given him as an observatory, a house in the city, and 3000 crowns a year for life. So his instruments were set up once more, students flocked to hear him and to receive work at his hands—among them a poor youth, John Kepler, to whom he was very kind, and who became, as you know, a still greater man than his master.

But the spirit of Tycho was broken, and though some good work was done at Prague—more observations made, and the Rudolphine tables begun—yet the hand of death was upon him. A painful disease seized him, attended with sleeplessness and temporary delirium, during the paroxysms of which he frequently exclaimed, *Ne frustra vixisse videar*. ("Oh that it may not appear that I have lived in vain!")

Quietly, however, at last, and surrounded by his friends and relatives, this fierce, passionate soul passed away, on the 24th of October, 1601.

His beloved instruments, which were almost a part of himself, were stored by Rudolph in a museum with scrupulous care, until the taking of Prague by the Elector Palatine's troops. In this disturbed time they got smashed, dispersed, and converted to other purposes. One thing only was saved—the great brass globe, which some thirty years after was recognized by a later king of Denmark as having belonged to Tycho, and deposited in the Library of the Academy of Sciences at Copenhagen, where I believe it is to this day.

The island of Huen was overrun by the Danish nobility, and nothing now remains of Uraniburg but a mound of earth and two pits.

As to the real work of Tycho, that has become immortal enough,—chiefly through the labours of his friend and scholar whose life we shall consider in the next lecture.

SUMMARY OF FACTS FOR LECTURE III

Life and work of Kepler. Kepler was born in December, 1571, at Weil in Würtemberg. Father an officer in the duke's army, mother something of a virago, both very poor. Kepler was utilized as a tavern pot-boy, but ultimately sent to a charity school, and thence to the University of Tübingen. Health extremely delicate; he was liable to violent attacks all his life. Studied mathematics, and accepted an astronomical lectureship at Graz as the first post which offered. Endeavoured to discover some connection between the number of the planets, their times of revolution, and their distances from the sun. Ultimately hit upon his fanciful regular-solid hypothesis, and published his first book in 1597. In 1599 was invited by Tycho to Prague, and there appointed Imperial mathematician, at a handsome but seldom paid salary. Observed the new star of 1604. Endeavoured to find the law of refraction of light from Vitellio's measurements, but failed. Analyzed Tycho's observations to find the true law of motion of Mars. After incredible labour, through innumerable wrong guesses, and six years of almost incessant calculation, he at length emerged in his two "laws"—discoveries which swept away all epicycles, deferents, equants, and other remnants of the Greek system, and ushered in the dawn of modern astronomy.

LAW I. *Planets move in ellipses, with the Sun in one focus.*

LAW II. *The radius vector (or line joining sun and planet) sweeps out equal areas in equal times.*

Published his second book containing these laws in 1609. Death of Rudolph in 1612, and subsequent increased misery and misfortune of Kepler. Ultimately discovered the connection between the times and distances of the planets for which he had been groping all his mature life, and announced it in 1618:—

LAW III. *The square of the time of revolution (or year) of each planet is proportional to the cube of its mean distance from the sun.*

The book in which this law was published ("On Celestial Harmonies") was dedicated to James of England. In 1620 had to intervene to protect his mother from being tortured for witchcraft. Accepted a professorship at Linz. Published the Rudolphine tables in 1627, embodying Tycho's observations and his own theory. Made a last effort to overcome his poverty by getting the arrears of his salary paid at Prague, but was unsuccessful, and, contracting brain fever on the journey, died in November, 1630, aged 59.

A man of keen imagination, indomitable perseverance, and uncompromising love of truth, Kepler overcame ill-health, poverty, and misfortune, and placed himself in the very highest rank of scientific men. His laws, so extraordinarily discovered, introduced order and simplicity into what else would have been a chaos of detailed observations; and they served as a secure basis for the splendid erection made on them by Newton.

Seven planets of the Ptolemaic system—
Moon, Mercury, Venus, Sun, Mars, Jupiter,
Saturn.
Six planets of the Copernican system—
Mercury, Venus, Earth, Mars, Jupiter,
Saturn.
The five regular solids, in appropriate order—
Octahedron, Icosahedron, Dodecahedron,
Tetrahedron, Cube.

Table illustrating Kepler's third law.

Planet.	Mean distance from Sun. D	Length of Year. T	Cube of the Distance. D^3	Square of the Time. T^2
Mercury	·3871	·24084	·05801	·05801
Venus	·7233	·61519	·37845	·37846
Earth	1·0000	1·0000	1·0000	1·0000
Mars	1·5237	1·8808	3·5375	3·5375
Jupiter	5·2028	11·862	140·83	140·70
Saturn	9·5388	29·457	867·92	867·70

The length of the earth's year is 365·256 days; its mean distance from the sun, taken above as unity, is 92,000,000 miles.

LECTURE III

KEPLER AND THE LAWS OF PLANETARY MOTION

It is difficult to imagine a stronger contrast between two men engaged in the same branch of science than exists between Tycho Brahé, the subject of last lecture, and Kepler, our subject on the present occasion.

The one, rich, noble, vigorous, passionate, strong in mechanical ingenuity and experimental skill, but not above the average in theoretical and mathematical power.

The other, poor, sickly, devoid of experimental gifts, and unfitted by nature for accurate observation, but strong almost beyond competition in speculative subtlety and innate mathematical perception.

The one is the complement of the other; and from the fact of their following each other so closely arose the most surprising benefits to science.

The outward life of Kepler is to a large extent a mere record of poverty and misfortune. I shall only sketch in its broad features, so that we may have more time to attend to his work.

He was born (so his biographer assures us) in longitude 29° 7', latitude 48° 54', on the 21st of December, 1571. His parents seem to have been of fair condition, but by reason, it is said, of his becoming surety for a friend, the father lost all his slender income, and was reduced to keeping a tavern. Young John Kepler was thereupon taken from school, and employed as pot-boy between the ages of nine and twelve. He was a sickly lad, subject to violent illnesses from the cradle, so that his life was frequently despaired of. Ultimately he was sent to a monastic school and thence to the University of Tübingen, where he graduated second on the list. Meanwhile home affairs had gone to rack and ruin. His father abandoned the home, and later died abroad. The mother quarrelled with all her relations, including her son John; who was therefore glad to get away as soon as possible.

All his connection with astronomy up to this time had been the hearing the Copernican theory expounded in University lectures, and defending it in a college debating society.

An astronomical lectureship at Graz happening to offer itself, he was urged to take it, and agreed to do so, though stipulating that it should not debar him from some more brilliant profession when there was a chance.

For astronomy in those days seems to have ranked as a minor science, like mineralogy or meteorology now. It had little of the special dignity with which the labours of Kepler himself were destined so greatly to aid in endowing it.

Well, he speedily became a thorough Copernican, and as he had a most singularly restless and inquisitive mind, full of appreciation of everything relating to number and magnitude—was a born speculator and thinker just as Mozart was a born musician, or Bidder a born calculator—he was agitated by questions such as these: Why are there exactly six planets? Is there any connection between their orbital distances, or between their orbits and the times of describing them? These things tormented him, and he thought about them day and night. It is characteristic of the spirit of the times—this questioning why there should be six planets. Nowadays, we should simply record the fact and look out for a seventh. Then, some occult property of the number six was groped for, such as that it was equal to 1 + 2 + 3 and likewise equal to 1 × 2 × 3, and so on. Many fine reasons had been given for the seven planets of the Ptolemaic system (see, for instance, p. 106), but for the six planets of the Copernican system the reasons were not so cogent.

Again, with respect to their successive distances from the sun, some law would seem to regulate their distance, but it was not known. (Parenthetically I may remark that it is not known even now: a crude empirical statement known as Bode's law—see page 294—is all that has been discovered.)

Once more, the further the planet the slower it moved; there seemed to be some law connecting speed and distance. This also Kepler made continual attempts to discover.

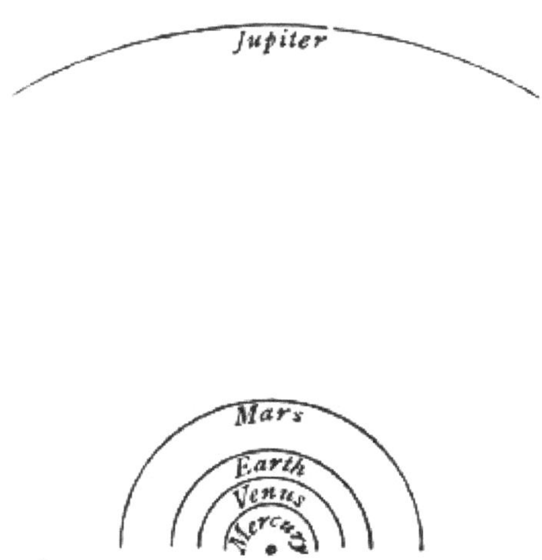

FIG. 26.—Orbits of some of the planets drawn to scale: showing the gap between Mars and Jupiter.

One of his ideas concerning the law of the successive distances was based on the inscription of a triangle in a circle. If you inscribe in a circle a large number of equilateral triangles, they envelop another circle bearing a definite ratio to the first: these might do for the orbits of two planets (see Fig. 27). Then try inscribing and circumscribing squares, hexagons, and other figures, and see if the circles thus defined would correspond to the several planetary orbits. But they would not give any satisfactory result. Brooding over this disappointment, the idea of trying solid figures suddenly strikes him. "What have plane figures to do with the celestial orbits?" he cries out; "inscribe the regular solids." And then—brilliant idea—he remembers that there are but five. Euclid had shown that there could be only five regular solids.[4] The number evidently corresponds to the gaps between the six planets. The reason of there being only six seems to be attained. This coincidence assures him he is on the right track, and with great enthusiasm and hope he "represents the earth's orbit by a sphere as the norm and measure of all"; round it he circumscribes a dodecahedron, and puts another sphere round that, which is approximately the orbit of Mars; round that, again, a tetrahedron, the corners of which mark the sphere of the orbit of Jupiter; round that sphere, again, he places a cube, which roughly gives the orbit of Saturn.

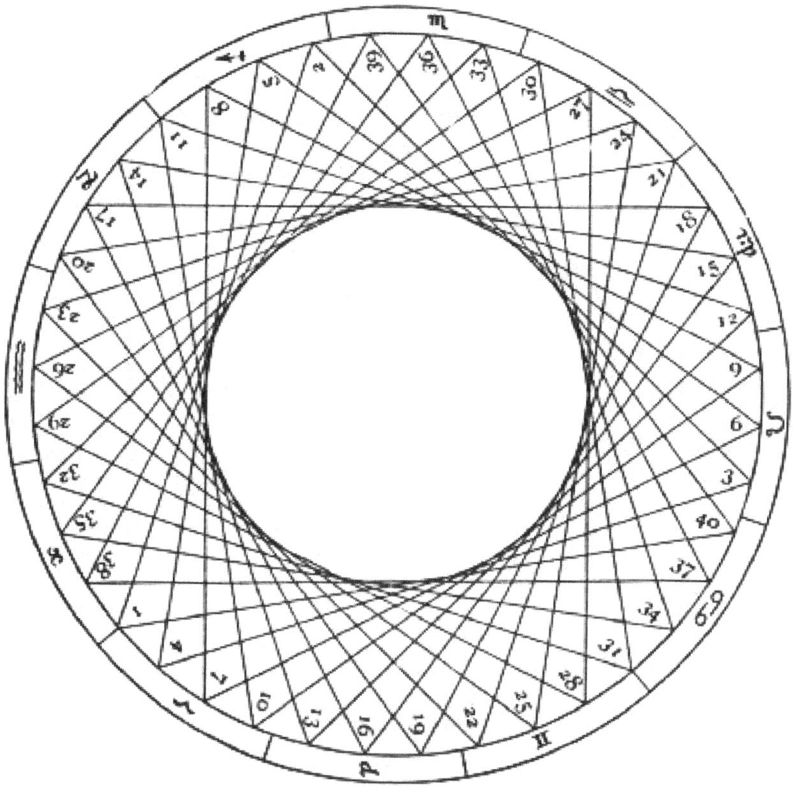

FIG. 27.—Many-sided polygon or approximate circle enveloped by straight lines, as for instance by a number of equilateral triangles.

On the other hand, he inscribes in the sphere of the earth's orbit an icosahedron; and inside the sphere determined by that, an octahedron; which figures he takes to inclose the spheres of Venus and of Mercury respectively.

The imagined discovery is purely fictitious and accidental. First of all, eight planets are now known; and secondly, their real distances agree only very approximately with Kepler's hypothesis.

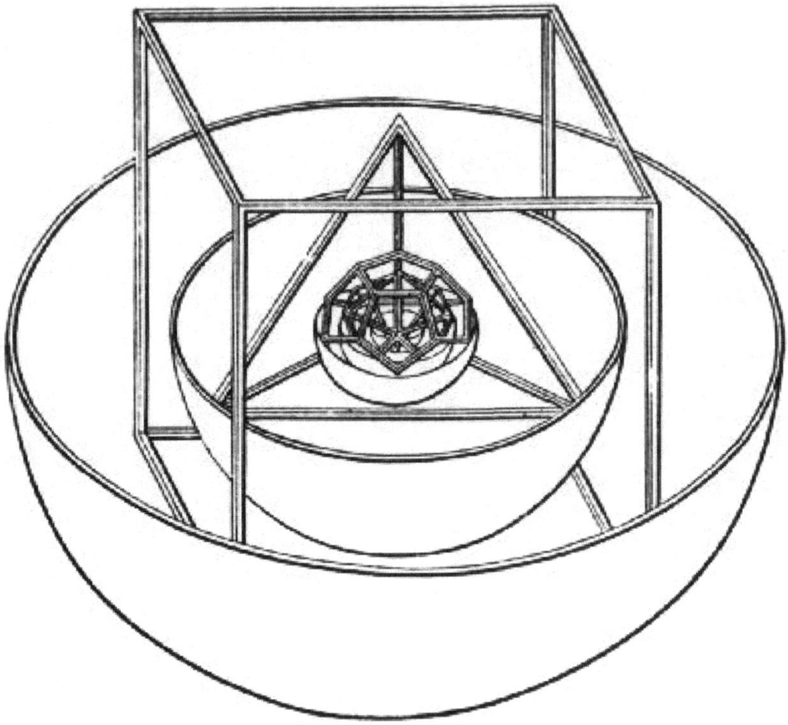

FIG. 28.—Frameworks with inscribed and circumscribed spheres, representing the five regular solids distributed as Kepler supposed them to be among the planetary orbits. (See "Summary" at beginning of this lecture, p. 57.)

Nevertheless, the idea gave him great delight. He says:—"The intense pleasure I have received from this discovery can never be told in words. I regretted no more the time wasted; I tired of no labour; I shunned no toil of reckoning, days and nights spent in calculation, until I could see whether my hypothesis would agree with the orbits of Copernicus, or whether my joy was to vanish into air."

He then went on to speculate as to the cause of the planets' motion. The old idea was that they were carried round by angels or celestial intelligences. Kepler tried to establish some propelling force emanating from the sun, like the spokes of a windmill.

This first book of his brought him into notice, and served as an introduction to Tycho and to Galileo.

Tycho Brahé was at this time at Prague under the patronage of the Emperor Rudolph; and as he was known to have by far the best planetary observations of any man living, Kepler wrote to him to know if he might come and examine them so as to perfect his theory.

Tycho immediately replied, "Come, not as a stranger, but as a very welcome friend; come and share in my observations with such instruments as I have with me, and as a dearly beloved associate." After this visit, Tycho wrote again, offering him the post of mathematical assistant, which after hesitation was accepted. Part of the hesitation Kepler expresses by saying that "for observations his sight was dull, and for mechanical operations his hand was awkward. He suffered much from weak eyes, and dare not expose himself to night air." In all this he was, of course, the antipodes of Tycho, but in mathematical skill he was greatly his superior.

On his way to Prague he was seized with one of his periodical illnesses, and all his means were exhausted by the time he could set forward again, so that he had to apply for help to Tycho.

It is clear, indeed, that for some time now he subsisted entirely on the bounty of Tycho, and he expresses himself most deeply grateful for all the kindness he received from that noble and distinguished man, the head of the scientific world at that date.

To illustrate Tycho's kindness and generosity, I must read you a letter written to him by Kepler. It seems that Kepler, on one of his absences from Prague, driven half mad with poverty and trouble, fell foul of Tycho, whom he thought to be behaving badly in money matters to him and his family, and wrote him a violent letter full of reproaches and insults. Tycho's secretary replied quietly enough, pointing out the groundlessness and ingratitude of the accusation.

Kepler repents instantly, and replies:—

"MOST NOBLE TYCHO," (these are the words of his letter), "how shall I enumerate or rightly estimate your benefits conferred on me? For two months you have liberally and gratuitously maintained me, and my whole family; you have provided for all my wishes; you have done me every possible kindness; you have communicated to me everything you hold most dear; no one, by word or deed, has intentionally injured me in anything; in short, not to your children, your wife, or yourself have you shown more indulgence than to me. This being so, as I am anxious to put on record, I cannot reflect without consternation that I should have been so given up by God to my own intemperance as to shut my eyes on all these benefits; that, instead of modest and respectful gratitude, I should indulge for three weeks in continual moroseness towards all your family, in headlong passion and the utmost insolence towards yourself, who possess so many claims on my veneration, from your noble family, your extraordinary learning, and distinguished reputation. Whatever I have said or written against the person, the fame, the honour, and the learning of your excellency; or whatever, in any other way, I have injuriously spoken or written (if they admit no other more favourable interpretation), as, to my grief, I have spoken and written many things, and more than I can remember; all and everything I recant, and freely and honestly declare and profess to be groundless, false, and incapable of proof."

Tycho accepted the apology thus heartily rendered, and the temporary breach was permanently healed.

In 1601, Kepler was appointed "Imperial mathematician," to assist Tycho in his calculations.

The Emperor Rudolph did a good piece of work in thus maintaining these two eminent men, but it is quite clear that it was as astrologers that he valued them; and all he cared for in the planetary motions was limited to their supposed effect on his own and his kingdom's destiny. He seems to have been politically a weak and superstitious prince, who was letting his kingdom get into hopeless confusion, and entangling himself in all manner of political complications. While Bohemia suffered, however, the world has benefited at his hands; and the tables upon which Tycho was now engaged are well called the Rudolphine tables.

These tables of planetary motion Tycho had always regarded as the main work of his life; but he died before they were finished, and on his death-bed he intrusted the completion of them to Kepler, who loyally undertook their charge.

The Imperial funds were by this time, however, so taxed by wars and other difficulties that the tables could only be proceeded with very slowly, a staff of calculators being out of the question. In fact, Kepler could not get even his own salary paid: he got orders, and promises, and drafts on estates for it; but when the time came for them to be honoured they were worthless, and he had no power to enforce his claims.

So everything but brooding had to be abandoned as too expensive, and he proceeded to study optics. He gave a very accurate explanation of the action of the human eye, and made many hypotheses, some of them shrewd and close to the mark, concerning the law of refraction of light in dense media: but though several minor points of interest turned up, nothing of the first magnitude came out of this long research.

The true law of refraction was discovered some years after by a Dutch professor, Willebrod Snell.

We must now devote a little time to the main work of Kepler's life. All the time he had been at Prague he had been making a severe study of the motion of the planet Mars, analyzing minutely Tycho's books of observations, in order to find out, if possible, the true theory of his motion. Aristotle had taught that circular motion was the only perfect and natural motion, and that the heavenly bodies therefore necessarily moved in circles.

So firmly had this idea become rooted in men's minds, that no one ever seems to have contemplated the possibility of its being false or meaningless.

When Hipparchus and others found that, as a matter of fact, the planets did *not* revolve in simple circles, they did not try other curves, as we should at once do now, but they tried combinations of circles, as we saw in Lecture I. The small circle carried by a bigger one was called an Epicycle. The carrying circle was called the Deferent. If for any reason the earth had to be placed out of the centre, the main planetary orbit was called an Excentric, and so on.

But although the planetary paths might be roughly represented by a combination of circles, their speeds could not, on the hypothesis of uniform motion in each circle round the earth as a fixed body. Hence was introduced the idea of an Equant, *i.e.* an arbitrary point, not the earth, about which the speed might be uniform. Copernicus, by making the sun the centre, had been able to simplify a good deal of this, and to abolish the equant.

But now that Kepler had the accurate observations of Tycho to refer to, he found immense difficulty in obtaining the true positions of the planets for long together on any such theory.

He specially attacked the motion of the planet Mars, because that was sufficiently rapid in its changes for a considerable collection of data to have accumulated with respect to it. He tried all manner of circular orbits for the earth and for Mars, placing them in all sorts of aspects with respect to the sun. The problem to be solved was to choose such an orbit and such a law of speed, for both the earth and Mars, that a line joining them, produced out to the stars, should always mark correctly the apparent position of Mars as seen from the earth. He had to arrange the size of the orbits that suited best, then the positions of their centres, both being supposed excentric with respect to the sun; but he could not get any such arrangement to work with uniform motion about the sun. So he reintroduced the equant, and thus had another variable at his disposal—in fact, two, for he had an equant for the earth and another for Mars, getting a pattern of the kind suggested in Fig. 29.

The equants might divide the line in any arbitrary ratio. All sorts of combinations had to be tried, the relative positions of the earth and Mars to be worked out for each, and compared with Tycho's recorded observations. It was easy to get them to agree for a short time, but sooner or later a discrepancy showed itself.

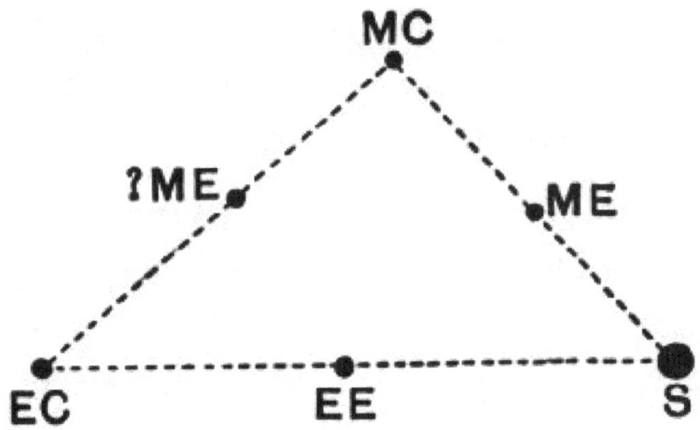

FIG. 29.—*S* represents the sun; *EC*, the centre of the earth's orbit, to be placed as best suited; *MC*, the same for Mars; *EE*, the earth's equant, or point about which the earth uniformly revolved (*i.e.* the point determining the law of speed about the sun), likewise to be placed anywhere, but supposed to be in the line joining *S* to *EC*; *ME*, the same thing for Mars; with *?ME* for an alternative hypothesis that perhaps Mars' equant was on line joining *EC* with *MC*.

I need not say that all these attempts and gropings, thus briefly summarized, entailed enormous labour, and required not only great pertinacity, but a most singularly constituted mind, that could thus continue groping in the dark without a possible ray of theory to illuminate its search. Grope he did, however, with unexampled diligence.

At length he hit upon a point that seemed nearly right. He thought he had found the truth; but no, before long the position of the planet, as calculated, and as recorded by Tycho, differed by eight minutes of arc, or about one-eighth of a degree. Could the observation be wrong by this small amount? No, he had known Tycho, and knew that he was never wrong eight minutes in an observation.

So he set out the whole weary way again, and said that with those eight minutes he would yet find out the law of the universe. He proceeded to see if by making the planet librate, or the plane of its orbit tilt up and down, anything could be done. He was rewarded by finding that at any rate the plane of the orbit did not tilt up and down: it was fixed, and this was a simplification on Copernicus's theory. It is not an absolute fixture, but the changes are very small (see Laplace, page 266).

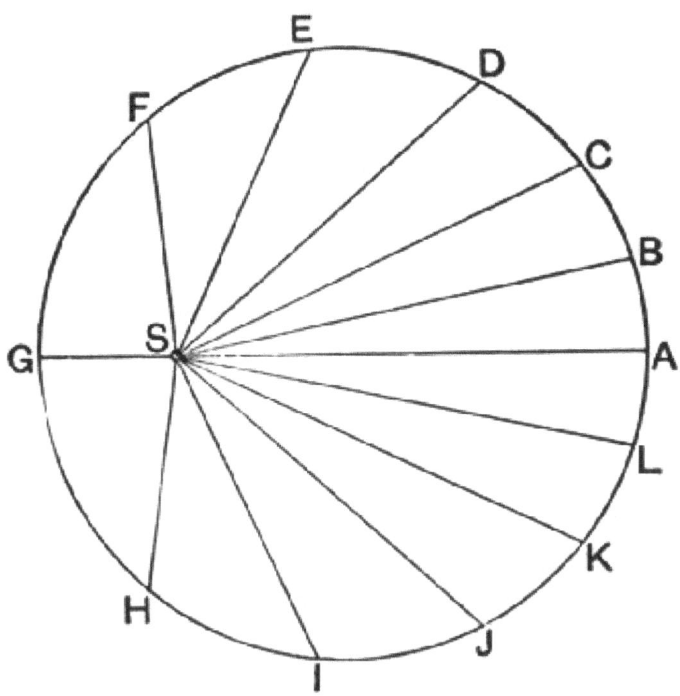

FIG. 30.—Excentric circle supposed to be divided into equal areas. The sun, *S*, being placed at a selected point, it was possible to represent the varying speed of a planet by saying that it moved from *A* to *B*, from *B* to *C*, and so on, in equal times.

At last he thought of giving up the idea of *uniform* circular motion, and of trying *varying* circular motion, say inversely as its distance from the sun. To simplify calculation, he divided the orbit into triangles, and tried if making the triangles equal would do. A great piece of luck, they did beautifully: the rate of description of areas (not arcs) is uniform. Over this discovery he greatly rejoices. He feels as though he had been carrying on a war against the planet and had triumphed; but his gratulation was premature. Before long fresh little errors appeared, and grew in importance. Thus he announces it himself:—

"While thus triumphing over Mars, and preparing for him, as for one already vanquished, tabular prisons and equated excentric fetters, it is buzzed here and there that the victory is vain, and that the war is raging anew as violently as before. For the enemy left at home a despised captive has burst all the chains of the equations, and broken forth from the prisons of the tables."

Still, a part of the truth had been gained, and was not to be abandoned any more. The law of speed was fixed: that which is now known as his second law. But what about the shape of the orbit—Was it after all possible that Aristotle, and every philosopher since Aristotle, had been wrong? that circular motion was not the perfect and natural motion, but that planets might move in some other closed curve?

Suppose he tried an oval. Well, there are a great variety of ovals, and several were tried: with the result that they could be made to answer better than a circle, but still were not right.

Now, however, the geometrical and mathematical difficulties of calculation, which before had been tedious and oppressive, threatened to become overwhelming; and it is with a rising sense of despondency that Kepler sees his six years' unremitting labour leading deeper and deeper into complication.

One most disheartening circumstance appeared, viz. that when he made the circuit oval his law of equable description of areas broke down. That seemed to require the circular orbit, and yet no circular orbit was quite accurate.

While thinking and pondering for weeks and months over this new dilemma and complication of difficulties, till his brain reeled, an accidental ray of light broke upon him in a way not now intelligible, or barely intelligible. Half the extreme breadth intercepted between the circle and oval was 429⁄100,000 of the radius, and he remembered that the "optical inequality" of Mars was also about 429⁄100,000. This coincidence, in his own words, woke him out of sleep; and for some reason or other impelled him instantly to try making the planet oscillate in the diameter of its epicycle instead of revolve round it—a singular idea, but Copernicus had had a similar one to explain the motions of Mercury.

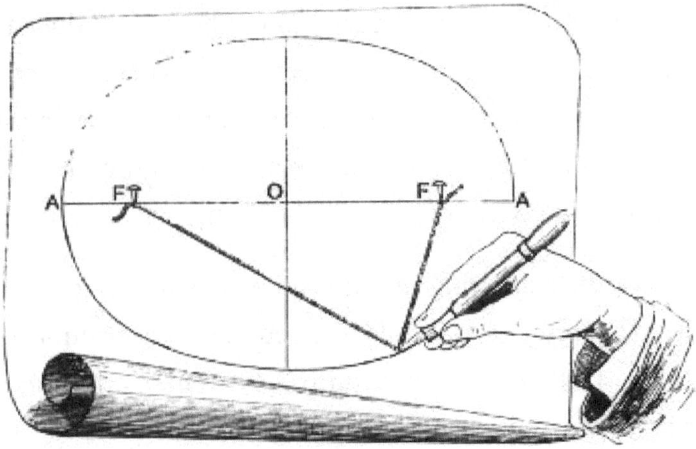

FIG. 31.—Mode of drawing an ellipse. The two pins *F* are the foci.

Away he started through his calculations again. A long course of work night and day was rewarded by finding that he was now able to hit off the motions better than before; but what a singularly complicated motion it was. Could it be expressed no more simply? Yes, the curve so described by the planet is a comparatively simple one: it is a special kind of oval—the ellipse. Strange that he had not thought of it before. It was a famous curve, for the Greek geometers had studied it as one of the sections of a cone, but it was not so well known in Kepler's time. The fact that the planets move in it has raised it to the first importance, and it is familiar enough to us now. But did it satisfy the law of speed? Could the rate of description of areas be uniform with it? Well, he tried the ellipse, and to his inexpressible delight he found that it did satisfy the condition of equable description of areas, if the sun was in one focus. So, moving the planet in a selected ellipse, with the sun in one focus, at a speed given by the equable area description, its position agreed with Tycho's observations within the limits of the error of experiment.

Mars was finally conquered, and remains in his prison-house to this day. The orbit was found.

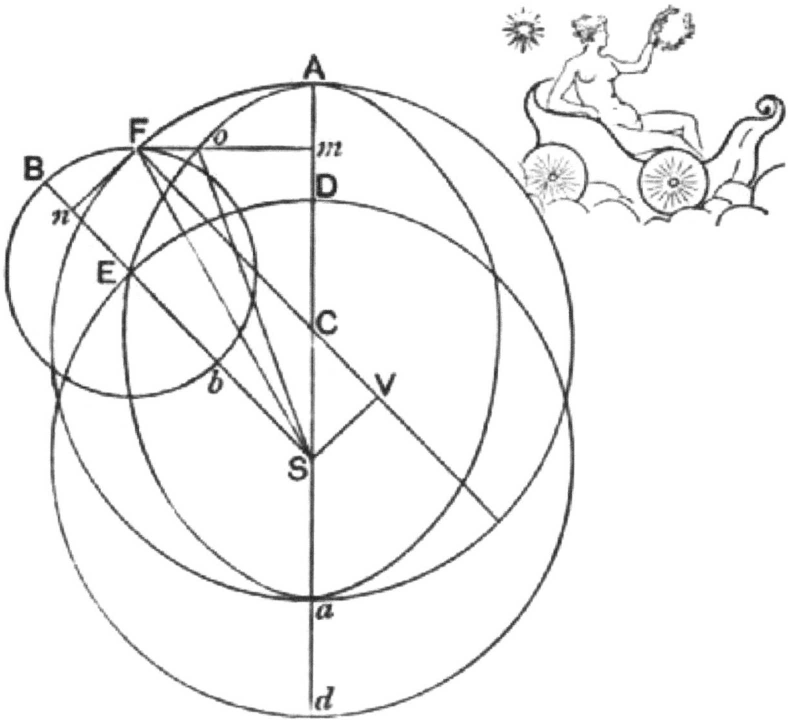

FIG. 32.

In a paroxysm of delight Kepler celebrates his victory by a triumphant figure, sketched actually on his geometrical diagram—the diagram which proves that the law of equable description of areas can hold good with an ellipse. The above is a tracing of it.

Such is a crude and bald sketch of the steps by which Kepler rose to his great generalizations—the two laws which have immortalized his name.

All the complications of epicycle, equant, deferent, excentric, and the like, were swept at once away, and an orbit of striking and beautiful properties substituted. Well might he be called, as he was, "the legislator," or law interpreter, "of the heavens."

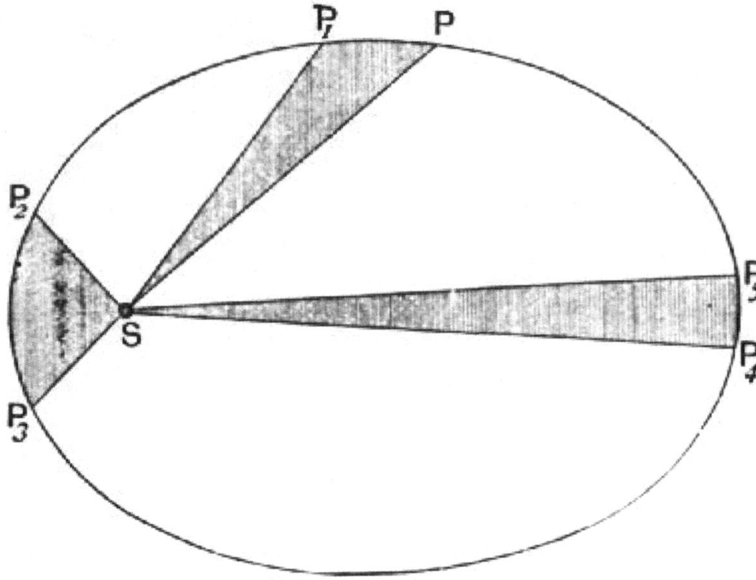

FIG. 33.—If *S* is the sun, a planet or comet moves from *P* to P_1, from P_2 to P_3, and from P_4 to P_5 in the same time; if the shaded areas are equal.

He concludes his book on the motions of Mars with a half comic appeal to the Emperor to provide him with the sinews of war for an attack on Mars's relations—father Jupiter, brother Mercury, and the rest—but the death of his unhappy patron in 1612 put an end to all these schemes, and reduced Kepler to the utmost misery. While at Prague his salary was in continual arrear, and it was with difficulty that he could provide sustenance for his family. He had been there eleven years, but they had been hard years of poverty, and he could leave without regret were it not that he should have to leave Tycho's instruments and observations behind him. While he was hesitating what best to do, and reduced to the verge of despair, his wife, who had long been suffering from low spirits and despondency, and his three children, were taken ill; one of the sons died of small-pox, and the wife eleven days after of low fever and epilepsy. No money could be got at Prague, so after a short time he accepted a professorship at Linz, and withdrew with his two quite young remaining children.

He provided for himself now partly by publishing a prophesying almanack, a sort of Zadkiel arrangement—a thing which he despised, but the support of which he could not afford to do without. He is continually attacking and throwing sarcasm at astrology, but it was the only thing for which people would pay him, and on it after a fashion he lived. We do not find that his circumstances were ever prosperous, and though 8,000 crowns were due to him from Bohemia he could not manage to get them paid.

About this time occurred a singular interruption to his work. His old mother, of whose fierce temper something has already been indicated, had been engaged in a law-suit for some years near their old home in Würtemberg. A change of judge having in process of time occurred, the defendant saw his way to turn the tables on the old lady by accusing her of sorcery. She was sent to prison, and condemned to the torture, with the usual intelligent idea of extracting a "voluntary" confession. Kepler had to hurry from Linz to interpose. He succeeded in saving her from the torture, but she remained in prison for a year or so. Her spirit, however, was unbroken, for no sooner was she released than she commenced a fresh action against her accuser. But fresh trouble was averted by the death of the poor old dame at the age of nearly eighty.

This narration renders the unflagging energy shown by her son in his mathematical wrestlings less surprising.

Interspersed with these domestic troubles, and with harassing and unsuccessful attempts to get his rights, he still brooded over his old problem of some possible connection between the distances of the planets from the sun and their times of revolution, *i.e.* the length of their years.

It might well have been that there was no connection, that it was purely imaginary, like his old idea of the law of the successive distances of the planets, and like so many others of the guesses and fancies which he entertained and spent his energies in probing. But fortunately this time there was a connection, and he lived to have the joy of discovering it.

The connection is this, that if one compares the distance of the different planets from the sun with the length of time they take to go round him, the cube of the respective distances is proportional to the square of the corresponding times. In other words, the ratio of r^3 to T^2 for every planet is the same. Or, again, the length of a planet's year depends on the $3/2$th power of its distance from the sun. Or, once more, the speed of each planet in its orbit is as the inverse square-root of its distance from the sun. The product of the distance into the square of the speed is the same for each planet.

This (however stated) is called Kepler's third law. It welds the planets together, and shows them to be one system. His rapture on detecting the law was unbounded, and he breaks out into an exulting rhapsody:—

"What I prophesied two-and-twenty years ago, as soon as I discovered the five solids among the heavenly orbits—what I firmly believed long before I had seen Ptolemy's *Harmonies*—what I had promised my friends in the title of this book, which I named before I was sure of my discovery—what sixteen years ago, I urged as a thing to be sought—that for which I joined Tycho Brahé, for which I settled in Prague, for which I have devoted the best part of my life to astronomical contemplations, at length I have brought to light, and recognized its truth beyond my most sanguine expectations. It is not eighteen months since I got the first glimpse of light, three months since the dawn, very few days since the unveiled sun, most admirable to gaze upon, burst upon me. Nothing holds me; I will indulge my sacred fury; I will triumph over mankind by the honest confession that I have stolen the golden vases of the Egyptians to build up a tabernacle for my God far away from the confines of Egypt. If you forgive me, I rejoice; if you are angry, I can bear it; the die is cast, the book is written, to be read either now or by posterity, I care not which; it may well wait a century for a reader, as God has waited six thousand years for an observer."

Soon after this great work his third book appeared: it was an epitome of the Copernican theory, a clear and fairly popular exposition of it, which had the honour of being at once suppressed and placed on the list of books prohibited by the Church, side by side with the work of Copernicus himself, *De Revolutionibus Orbium Cœlestium*.

This honour, however, gave Kepler no satisfaction—it rather occasioned him dismay, especially as it deprived him of all pecuniary benefit, and made it almost impossible for him to get a publisher to undertake another book.

Still he worked on at the Rudolphine tables of Tycho, and ultimately, with some small help from Vienna, completed them; but he could not get the means to print them. He applied to the Court till he was sick of applying: they lay idle four years. At last he determined to pay for the type himself. What he paid it with, God knows, but he did pay it, and he did bring out the tables, and so was faithful to the behest of his friend.

This great publication marks an era in astronomy. They were the first really accurate tables which navigators ever possessed; they were the precursors of our present *Nautical Almanack.*

After this, the Grand Duke of Tuscany sent Kepler a golden chain, which is interesting inasmuch as it must really have come from Galileo, who was in high favour at the Italian Court at this time.

Once more Kepler made a determined attempt to get his arrears of salary paid, and rescue himself and family from their bitter poverty. He travelled to Prague on purpose, attended the imperial meeting, and pleaded his own cause, but it was all fruitless; and exhausted by the journey, weakened by over-study, and disheartened by the failure, he caught a fever, and died in his fifty-ninth year. His body was buried at Ratisbon, and a century ago a proposal was made to erect a marble monument to his memory, but nothing was done. It matters little one way or the other whether Germany, having almost refused him bread during his life, should, a century and a half after his death, offer him a stone.

FIG. 34.—Portrait of Kepler, older.

The contiguity of the lives of Kepler and Tycho furnishes a moral too obvious to need pointing out. What Kepler might have achieved had he been relieved of those ghastly struggles for subsistence one cannot tell, but this much is clear, that had Tycho been subjected to the same misfortune, instead of being born rich and being assisted by generous and enlightened patrons, he could have accomplished very little. His instruments, his observatory—the tools by which he did his work—would have been impossible for him. Frederick and Sophia of Denmark, and Rudolph of Bohemia, are therefore to be remembered as co-workers with him.

Kepler, with his ill-health and inferior physical energy, was unable to command the like advantages. Much, nevertheless, he did; more one cannot but feel he might have done had he been properly helped. Besides, the world would have been free from the reproach of accepting the fruits of his bright genius while condemning the worker to a life of misery, relieved only by the beauty of his own thoughts and the ecstasy awakened in him by the harmony and precision of Nature.

Concerning the method of Kepler, the mode by which he made his discoveries, we must remember that he gives us an account of all the steps, unsuccessful as well as successful, by which he travelled. He maps out his route like a traveller. In fact he compares himself to Columbus or Magellan, voyaging into unknown lands, and recording his wandering route. This being remembered, it will be found that his methods do not differ so utterly from those used by other philosophers in like case. His imagination was perhaps more luxuriant and was allowed freer play than most men's, but it was nevertheless always controlled by rigid examination and comparison of hypotheses with fact.

Brewster says of him:—"Ardent, restless, burning to distinguish himself by discovery, he attempted everything; and once having obtained a glimpse of a clue, no labour was too hard in following or verifying it. A few of his attempts succeeded—a multitude failed. Those which failed seem to us now fanciful, those which succeeded appear to us sublime. But his methods were the same. When in search of what really existed he sometimes found it; when in pursuit of a chimæra he could not but fail; but in either case he displayed the same great qualities, and that obstinate perseverance which must conquer all difficulties except those really insurmountable."

To realize what he did for astronomy, it is necessary for us now to consider some science still in its infancy. Astronomy is so clear and so thoroughly explored now, that it is difficult to put oneself into a contemporary attitude. But take some other science still barely developed: meteorology, for instance. The science of the weather, the succession of winds and rain, sunshine and frost, clouds and fog, is now very much in the condition of astronomy before Kepler.

We have passed through the stage of ascribing atmospheric disturbances—thunderstorms, cyclones, earthquakes, and the like—to supernatural agency; we have had our Copernican era: not perhaps brought about by a single individual, but still achieved. Something of the laws of cyclone and anticyclone are known, and rude weather predictions across the Atlantic are roughly possible. Barometers and thermometers and anemometers, and all their tribe, represent the astronomical instruments in the island of Huen; and our numerous meteorological observatories, with their continual record of events, represent the work of Tycho Brahé.

Observation is heaped on observation; tables are compiled; volumes are filled with data; the hours of sunshine are recorded, the fall of rain, the moisture in the air, the kind of clouds, the temperature—millions of facts; but where is the Kepler to study and brood over them? Where is the man to spend his life in evolving the beginnings of law and order from the midst of all this chaos?

Perhaps as a man he may not come, but his era will come. Through this stage the science must pass, ere it is ready for the commanding intellect of a Newton.

But what a work it will be for the man, whoever he be that undertakes it—a fearful monotonous grind of calculation, hypothesis, hypothesis, calculation, a desperate and groping endeavour to reconcile theories with facts.

A life of such labour, crowned by three brilliant discoveries, the world owes (and too late recognizes its obligation) to the harshly treated German genius, Kepler.

SUMMARY OF FACTS FOR LECTURES IV AND V

In 1564, Michael Angelo died and Galileo was born; in 1642, Galileo died and Newton was born. Milton lived from 1608 to 1674.

For teaching the plurality of worlds, with other heterodox doctrines, and refusing to recant, Bruno, after six years' imprisonment in Rome, was burnt at the stake on the 16th of

February, 1600 A.D. A "natural" death in the dungeons of the Inquisition saved Antonio de Dominis, the explainer of the rainbow, from the same fate, but his body and books were publicly burned at Rome in 1624.

The persecution of Galileo began in 1615, became intense in 1632, and so lasted till his death and after.

Galileo Galilei, eldest son of Vincenzo de Bonajuti de Galilei, a noble Florentine, was born at Pisa, 18th of February, 1564. At the age of 17 was sent to the University of Pisa to study medicine. Observed the swing of a pendulum and applied it to count pulse-beats. Read Euclid and Archimedes, and could be kept at medicine no more. At 26 was appointed Lecturer in Mathematics at Pisa. Read Bruno and became smitten with the Copernican theory. Controverted the Aristotelians concerning falling bodies, at Pisa. Hence became unpopular and accepted a chair at Padua, 1592. Invented a thermometer. Wrote on astronomy, adopting the Ptolemaic system provisionally, and so opened up a correspondence with Kepler, with whom he formed a friendship. Lectured on the new star of 1604, and publicly renounced the old systems of astronomy. Invented a calculating compass or "Gunter's scale." In 1609 invented a telescope, after hearing of a Dutch optician's discovery. Invented the microscope soon after. Rapidly completed a better telescope and began a survey of the heavens. On the 8th of January, 1610, discovered Jupiter's satellites. Observed the mountains in the moon, and roughly measured their height.

Explained the visibility of the new moon by *earth-shine*. Was invited to the Grand Ducal Court of Tuscany by Cosmo de Medici, and appointed philosopher to that personage. Discovered innumerable new stars, and the nebulæ. Observed a triple appearance of Saturn. Discovered the phases of Venus predicted by Copernicus, and spots on the sun. Wrote on floating bodies. Tried to get his satellites utilized for determining longitude at sea.

Went to Rome to defend the Copernican system, then under official discussion, and as a result was formally forbidden ever to teach it. On the accession of Pope Urban VIII. in 1623, Galileo again visited Rome to pay his respects, and was well received. In 1632 appeared his "Dialogues" on the Ptolemaic and Copernican systems. Summoned to Rome, practically imprisoned, and "rigorously questioned." Was made to recant 22nd of June, 1633. Forbidden evermore to publish anything, or to teach, or receive friends. Retired to Arcetri in broken down health. Death of his favourite daughter, Sister Maria Celeste. Wrote and meditated on the laws of motion. Discovered the moon's libration. In 1637 he became blind. The rigour was then slightly relaxed and many visited him: among them John Milton. Died 8th of January, 1642, aged 78. As a prisoner of the Inquisition his right to make a will or to be buried in consecrated ground was disputed. Many of his manuscripts were destroyed.

Galileo, besides being a singularly clear-headed thinker and experimental genius, was also something of a musician, a poet, and an artist. He was full of humour as well as of solid common-sense, and his literary style is brilliant. Of his scientific achievements those now reckoned most weighty, are the discovery of the Laws of Motion, and the laying of the foundations of Mechanics.

Particulars of Jupiter's Satellites,
Illustrating their obedience to Kepler's third law.

Satellite.	Diameter in miles.	Time of revolution in hours. (T)	Distance from Jupiter, in Jovian radii. (d)	T^2	d^3	$\frac{T^2}{d^3}$ which is practically constant.
No. 1.	2437	42·47	6·049	1803·7	221·44	8·149
No. 2.	2188	85·23	9·623	7264·1	891·11	8·152
No. 3.	3575	177·72	15·350	29488·	3916·8	8·153
No. 4.	3059	400·53	26·998	160426·	19679·	8·152

The diameter of Jupiter is 85,823 miles.

Falling Bodies.

Since all bodies fall at the same rate, except for the disturbing effect of the resistance of the air, a statement of their rates of fall is of interest. In one second a freely falling body near the earth is found to drop 16 feet. In two seconds it drops 64 feet altogether, viz. 16 feet in the first, and 48 feet in the next second; because at the beginning of every second after the first it has the accumulated velocity of preceding seconds. The height fallen by a dropped body is not proportional to the time simply, but to what is rather absurdly called the square of the time, *i.e.* the time multiplied by itself.

For instance, in 3 seconds it drops 9 × 16 = 144 feet; in 4 seconds 16 × 16, or 256 feet, and so on. The distances travelled in 1, 2, 3, 4,

&c., seconds by a body dropped from rest and not appreciably resisted by the air, are 1, 4, 9, 16, 25, &c., respectively, each multiplied by the constant 16 feet.

Another way of stating the law is to say that the heights travelled in successive seconds proceed in the proportion 1, 3, 5, 7, 9, &c.; again multiplied by 16 feet in each case.

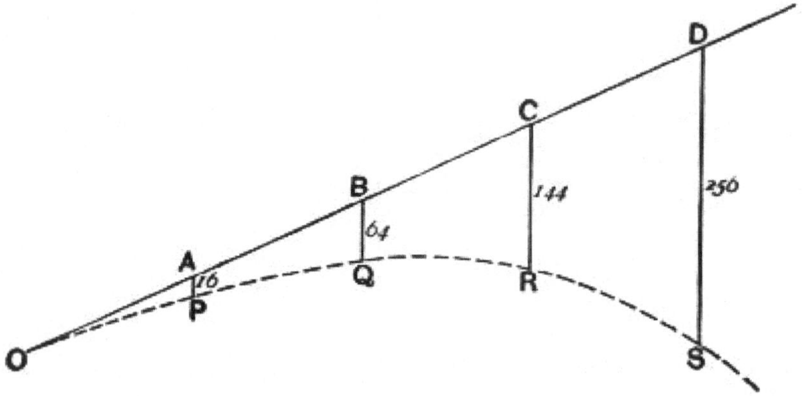

FIG. 35.—Curve described by a projectile, showing how it drops from the line of fire, *O D*, in successive seconds, the same distances *AP*, *BQ*, *CR*, &c., as are stated above for a dropped body.

All this was experimentally established by Galileo.

A body takes half a second to drop 4 feet; and a quarter of a second to drop 1 foot. The easiest way of estimating a quarter of a second with some accuracy is to drop a bullet one foot.

A bullet thrown or shot in any direction falls just as much as if merely dropped; but instead of falling from the starting-point it drops vertically from the line of fire. (See fig. 35).

The rate of fall depends on the intensity of gravity; if it could be doubled, a body would fall twice as far in the same time; but to make it fall a given distance in half the time the intensity of gravity would have to be quadrupled. At a place where the intensity of gravity is 1⁄3600 of what it is here, a body would fall as far in a minute as it now falls in a second. Such a place occurs at about the distance of the moon (*cf.* page 177).

The fact that the height fallen through is proportional to the square of the time proves that the attraction of the earth or the intensity of gravity is sensibly constant throughout ordinary small ranges. Over great distances of fall, gravity cannot be considered constant; so for things falling through great spaces the Galilean law of the square of the time does not hold.

The fact that things near the earth fall 16 feet in the first second proves that the intensity of ordinary terrestrial gravity is 32 British units of force per pound of matter.

The fact that all bodies fall at the same rate (when the resistance of the air is eliminated), proves that weight is proportional to mass; or more explicitly, that the gravitative attraction of the earth on matter near its surface depends on the amount of that matter, as estimated by its inertia, and on nothing else.

LECTURE IV

GALILEO AND THE INVENTION OF THE TELESCOPE

CONTEMPORARY with the life of Kepler, but overlapping it at both ends, comes the great and eventful life of Galileo Galilei,[5] a man whose influence on the development of human thought has been greater than that of any man whom we have yet considered, and upon whom, therefore, it is necessary for us, in order to carry out the plan of these lectures, to bestow much time. A man of great and wide culture, a so-called universal genius, it is as an experimental philosopher that he takes the first rank. In this capacity he must be placed alongside of Archimedes, and it is pretty certain that between the two there was no man of magnitude equal to either in experimental philosophy. It is perhaps too bold a speculation, but I venture to doubt whether in succeeding generations we find his equal in the domain of purely experimental science until we come to Faraday. Faraday was no doubt his superior, but I know of no other of whom the like can unhesitatingly be said. In mathematical and deductive science, of course, it is quite otherwise. Kepler, for instance, and many men before and since, have far excelled Galileo in mathematical skill and power, though at the same time his achievements in this department are by no means to be despised.

Born at Pisa three centuries ago, on the very day that Michael Angelo lay dying in Rome, he inherited from his father a noble name, cultivated tastes, a keen love of truth, and an impoverished patrimony. Vincenzo de Galilei, a descendant of the important

Bonajuti family, was himself a mathematician and a musician, and in a book of his still extant he declares himself in favour of free and open inquiry into scientific matters, unrestrained by the weight of authority and tradition.

In all probability the son imbibed these precepts: certainly he acted on them.

Vincenzo, having himself experienced the unremunerative character of scientific work, had a horror of his son's taking to it, especially as in his boyhood he was always constructing ingenious mechanical toys, and exhibiting other marks of precocity. So the son was destined for business—to be, in fact, a cloth-dealer. But he was to receive a good education first, and was sent to an excellent convent school.

Here he made rapid progress, and soon excelled in all branches of classics and literature. He delighted in poetry, and in later years wrote several essays on Dante, Tasso, and Ariosto, besides composing some tolerable poems himself. He played skilfully on several musical instruments, especially on the lute, of which indeed he became a master, and on which he solaced himself when quite an old man. Besides this he seems to have had some skill as an artist, which was useful afterwards in illustrating his discoveries, and to have had a fine sensibility as an art critic, for we find several eminent painters of that day acknowledging the value of the opinion of the young Galileo.

Perceiving all this display of ability, the father wisely came to the conclusion that the selling of woollen stuffs would hardly satisfy his aspirations for long, and that it was worth a sacrifice to send him to the University. So to the University of his native town he went, with the avowed object of studying medicine, that career seeming the most likely to be profitable. Old Vincenzo's horror of mathematics or science as a means of obtaining a livelihood is justified by the fact that while the University Professor of Medicine received 2,000 scudi a year, the Professor of Mathematics had only 60, that is £13 a year, or 7½*d.* a day.

So the son had been kept properly ignorant of such poverty-stricken subjects, and to study medicine he went.

But his natural bent showed itself even here. For praying one day in the Cathedral, like a good Catholic as he was all his life, his attention was arrested by the great lamp which, after lighting it, the verger had left swinging to and fro. Galileo proceeded to time its swings by the only watch he possessed—viz., his own pulse. He noticed that the time of swing remained as near as he could tell the same, notwithstanding the fact that the swings were getting smaller and smaller.

By subsequent experiment he verified the law, and the isochronism of the pendulum was discovered. An immensely important practical discovery this, for upon it all modern clocks are based; and Huyghens soon applied it to the astronomical clock, which up to that time had been a crude and quite untrustworthy instrument.

The best clock which Tycho Brahé could get for his observatory was inferior to one that may now be purchased for a few shillings; and this change is owing to the discovery of the pendulum by Galileo. Not that he applied it to clocks; he was not thinking of astronomy, he was thinking of medicine, and wanted to count people's pulses. The pendulum served; and "pulsilogies," as they were called, were thus introduced to and used by medical practitioners.

The Tuscan Court came to Pisa for the summer months, for it was then a seaside place, and among the suite was Ostillio Ricci, a distinguished mathematician and old friend of the Galileo family. The youth visited him, and one day, it is said, heard a lesson in Euclid being given by Ricci to the pages while he stood outside the door entranced. Anyhow he implored Ricci to help him into some knowledge of mathematics, and the old man willingly consented. So he mastered Euclid and passed on to Archimedes, for whom he acquired a great veneration.

His father soon heard of this obnoxious proclivity, and did what he could to divert him back to medicine again. But it was no use. Underneath his Galen and Hippocrates were secreted copies of Euclid and Archimedes, to be studied at every available opportunity. Old Vincenzo perceived the bent of genius to be too strong for him, and at last gave way.

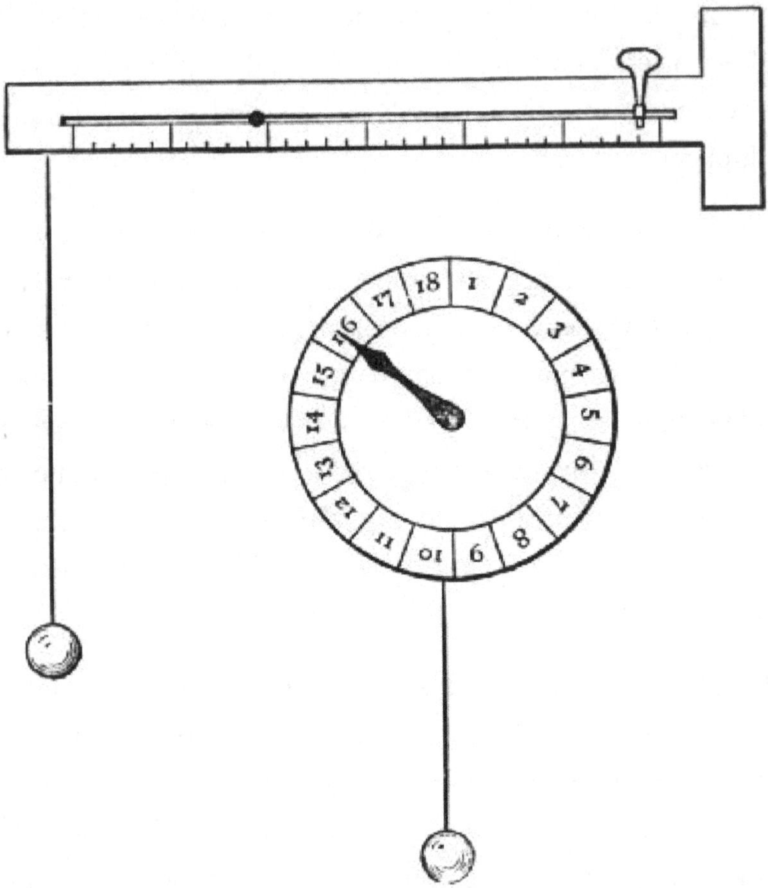

FIG. 36.—Two forms of pulsilogy. The string is wound up till the swinging weight keeps time with the pulse, and the position of a bead or of an index connected with the string is then read on a scale or dial.

With prodigious rapidity the released philosopher now assimilated the elements of mathematics and physics, and at twenty-six we find him appointed for three years to the University Chair of Mathematics, and enjoying the paternally dreaded stipend of 7½*d.* a day.

Now it was that he pondered over the laws of falling bodies. He verified, by experiment, the fact that the velocity acquired by falling down any slope of given height was independent of the angle of slope. Also, that the height fallen through was proportional to the square of the time.

Another thing he found experimentally was that all bodies, heavy and light, fell at the same rate, striking the ground at the same time.[6]

Now this was clean contrary to what he had been taught. The physics of those days were a simple reproduction of statements in old books. Aristotle had asserted certain things to be true, and these were universally believed. No one thought of trying the thing to see if it really were so. The idea of making an experiment would have savoured of impiety, because it seemed to tend towards scepticism, and cast a doubt on a reverend authority.

Young Galileo, with all the energy and imprudence of youth (what a blessing that youth has a little imprudence and disregard of consequences in pursuing a high ideal!), as soon as he perceived that his instructors were wrong on the subject of falling bodies, instantly informed them of the fact. Whether he expected them to be pleased or not is a question. Anyhow, they were not pleased, but were much annoyed by his impertinent arrogance.

It is, perhaps, difficult for us now to appreciate precisely their position. These doctrines of antiquity, which had come down hoary with age, and the discovery of which had reawakened learning and quickened intellectual life, were accepted less as a science or a philosophy, than as a religion. Had they regarded Aristotle as a verbally inspired writer, they could not have received his statements with more unhesitating conviction. In any dispute as to a question of fact, such as the one before us concerning the laws of falling bodies, their method was not to make an experiment, but to turn over the pages of Aristotle; and he who could quote chapter and verse of this great writer was held to settle the question and raise it above the reach of controversy.

It is very necessary for us to realize this state of things clearly, because otherwise the attitude of the learned of those days towards every new discovery seems stupid and almost insane. They had a crystallized system of truth, perfect, symmetrical—it wanted no novelty, no additions; every addition or growth was an imperfection, an excrescence, a deformity. Progress was unnecessary and undesired. The Church had a rigid system of dogma, which must be accepted in its entirety on pain of being treated as a heretic. Philosophers had a cast-iron system of truth to match—a system founded upon Aristotle—and so interwoven with the great theological dogmas that to question one was almost equivalent to casting doubt upon the other.

In such an atmosphere true science was impossible. The life-blood of science is growth, expansion, freedom, development. Before it could appear it must throw off these old shackles of centuries. It must burst its old skin, and emerge, worn with the struggle, weakly and unprotected, but free and able to grow and to expand. The conflict was inevitable, and it was severe. Is it over yet? I fear not quite, though so nearly as to disturb science hardly at all. Then it was different; it was terrible. Honour to the men who bore the first shock of the battle!

Now Aristotle had said that bodies fell at rates depending on their weight.

A 5 lb. weight would fall five times as quick as a 1 lb. weight; a 50 lb. weight fifty times as quick, and so on.

Why he said so nobody knows. He cannot have tried. He was not above trying experiments, like his smaller disciples; but probably it never occurred to him to doubt the fact. It seems so natural that a heavy body should fall quicker than a light one; and perhaps he thought of a stone and a feather, and was satisfied.

Galileo, however, asserted that the weight did not matter a bit, that everything fell at the same rate (even a stone and a feather, but for the resistance of the air), and would reach the ground in the same time.

And he was not content to be pooh-poohed and snubbed. He knew he was right, and he was determined to make every one see the facts as he saw them. So one morning, before the assembled University, he ascended the famous leaning tower, taking with him a 100 lb. shot and a 1 lb. shot. He balanced them on the edge of the tower, and let them drop together. Together they fell, and together they struck the ground.

The simultaneous clang of those two weights sounded the death-knell of the old system of philosophy, and heralded the birth of the new.

But was the change sudden? Were his opponents convinced? Not a jot. Though they had seen with their eyes, and heard with their ears, the full light of heaven shining upon them, they went back muttering and discontented to their musty old volumes and their garrets, there to invent occult reasons for denying the validity of the observation, and for referring it to some unknown disturbing cause.

They saw that if they gave way on this one point they would be letting go their anchorage, and henceforward would be liable to drift along with the tide, not knowing whither. They dared not do this. No; they *must* cling to the old traditions; they could not cast away their rotting ropes and sail out on to the free ocean of God's truth in a spirit of fearless faith.

FIG. 37.—Tower of Pisa.

Yet they had received a shock: as by a breath of fresh salt breeze and a dash of spray in their faces, they had been awakened out of their comfortable lethargy. They felt the approach of a new era.

Yes, it was a shock; and they hated the young Galileo for giving it them—hated him with the sullen hatred of men who fight for a lost and dying cause.

We need scarcely blame these men; at least we need not blame them overmuch. To say that they acted as they did is to say that they were human, were narrow-minded, and were the apostles of a lost cause. But *they* could not know this; *they* had no experience of the past to guide them; the conditions under which they found themselves were novel, and had to be met for the first time. Conduct which was excusable then would be unpardonable now, in the light of all this experience to guide us. Are there any now who practically repeat their error, and resist new truth? who cling to any old anchorage of dogma, and refuse to rise with the tide of advancing knowledge? There may be some even now.

Well, the unpopularity of Galileo smouldered for a time, until, by another noble imprudence, he managed to offend a semi-royal personage, Giovanni de Medici, by giving his real opinion, when consulted, about a machine which de Medici had invented for cleaning out the harbour of Leghorn. He said it was as useless as it in fact turned out to be. Through the influence of the mortified inventor he lost favour at Court; and his enemies took advantage of the fact to render his chair untenable. He resigned before his three years were up, and retired to Florence.

His father at this time died, and the family were left in narrow circumstances. He had a brother and three sisters to provide for.

He was offered a professorship at Padua for six years by the Senate of Venice, and willingly accepted it.

Now began a very successful career. His introductory address was marked by brilliant eloquence, and his lectures soon acquired fame. He wrote for his pupils on the laws of motion, on fortifications, on sundials, on mechanics, and on the celestial globe: some of these papers are now lost, others have been printed during the present century.

Kepler sent him a copy of his new book, *Mysterium Cosmographicum*, and Galileo in thanking him for it writes him the following letter:—[7]

"I count myself happy, in the search after truth, to have so great an ally as yourself, and one who is so great a friend of the truth itself. It is really pitiful that there are so few who seek truth, and who do not pursue a perverse method of philosophising. But this is not the place to mourn over the miseries of our times, but to congratulate you on your splendid discoveries in confirmation of truth. I shall read your book to the end, sure of finding much that is excellent in it. I shall do so with the more pleasure, because *I have been for many years an adherent of the Copernican system*, and it explains to me the causes of many of the appearances of nature which are quite unintelligible on the commonly accepted hypothesis. *I have collected many arguments for the purpose of refuting the latter*; but I do not venture to bring them to the light of publicity, for fear of sharing the fate of our master, Copernicus, who, although he has earned immortal fame with some, yet with very many (so great is the number of fools) has become an object of ridicule and scorn. I should certainly venture to publish my speculations if there were more people like you. But this not being the case, I refrain from such an undertaking."

Kepler urged him to publish his arguments in favour of the Copernican theory, but he hesitated for the present, knowing that his declaration would be received with ridicule and opposition, and thinking it wiser to get rather more firmly seated in his chair before encountering the storm of controversy.

The six years passed away, and the Venetian Senate, anxious not to lose so bright an ornament, renewed his appointment for another six years at a largely increased salary.

Soon after this appeared a new star, the stella nova of 1604, not the one Tycho had seen—that was in 1572—but the same that Kepler was so much interested in.

Galileo gave a course of three lectures upon it to a great audience. At the first the theatre was over-crowded, so he had to adjourn to a hall holding 1000 persons. At the next he had to lecture in the open air.

He took occasion to rebuke his hearers for thronging to hear about an ephemeral novelty, while for the much more wonderful and important truths about the permanent stars and facts of nature they had but deaf ears.

But the main point he brought out concerning the new star was that it upset the received Aristotelian doctrine of the immutability of the heavens. According to that doctrine the heavens were unchangeable, perfect, subject neither to growth nor to decay. Here was a body, not a meteor but a real distant star, which had not been visible and which would shortly fade away again, but which meanwhile was brighter than Jupiter.

The staff of petrified professorial wisdom were annoyed at the appearance of the star, still more at Galileo's calling public attention to it; and controversy began at Padua. However, he accepted it; and now boldly threw down the gauntlet in favour of the Copernican theory, utterly repudiating the old Ptolemaic system which up to that time he had taught in the schools according to established custom.

The earth no longer the only world to which all else in the firmament were obsequious attendants, but a mere insignificant speck among the host of heaven! Man no longer the centre and cynosure of creation, but, as it were, an insect crawling on the surface of this little speck! All this not set down in crabbed Latin in dry folios for a few learned monks, as in Copernicus's time, but promulgated and argued in rich Italian, illustrated by analogy, by experiment, and with cultured wit; taught not to a few scholars here and there in musty libraries, but proclaimed in the vernacular to the whole populace with all the energy and enthusiasm of a recent convert and a master of language! Had a bombshell been exploded among the fossilized professors it had been less disturbing.

But there was worse in store for them.

A Dutch optician, Hans Lippershey by name, of Middleburg, had in his shop a curious toy, rigged up, it is said, by an apprentice, and made out of a couple of spectacle lenses, whereby, if one looked through it, the weather-cock of a neighbouring church spire was seen nearer and upside down.

The tale goes that the Marquis Spinola, happening to call at the shop, was struck with the toy and bought it. He showed it to Prince Maurice of Nassau, who thought of using it for military reconnoitring. All this is trivial. What is important is that some faint and inaccurate echo of this news found its way to Padua, and into the ears of Galileo.

The seed fell on good soil. All that night he sat up and pondered. He knew about lenses and magnifying glasses. He had read Kepler's theory of the eye, and had himself lectured on optics. Could he not hit on the device and make an instrument capable of bringing the heavenly bodies nearer? Who knew what marvels he might not so perceive! By morning he had some schemes ready to try, and one of them was successful. Singularly enough it was not the same plan as the Dutch optician's, it was another mode of achieving the same end.

He took an old small organ pipe, jammed a suitably chosen spectacle glass into either end, one convex the other concave, and behold, he had the half of a wretchedly bad opera glass capable of magnifying three times. It was better than the Dutchman's, however; it did not invert.

It is easy to understand the general principle of a telescope. A general knowledge of the common magnifying glass may be assumed. Roger Bacon knew about lenses; and the ancients often refer to them, though usually as burning glasses. The magnifying power of globes of water must have been noticed soon after the discovery of glass and the art of working it.

A magnifying glass is most simply thought of as an additional lens to the eye. The eye has a lens by which ordinary vision is accomplished, an extra glass lens strengthens it and enables objects to be seen nearer and therefore apparently bigger. But to apply a magnifying glass to distant objects is impossible. In order to magnify distant objects, another function of lenses has also to be employed, viz., their power of forming real images, the power on which their use as burning-glasses depends: for the best focus is an image of the sun. Although the object itself is inaccessible, the image of it is by no means so, and to the image a magnifier can be applied. This is exactly what is done in the telescope; the object glass or large lens forms an image, which is then looked at through a magnifying glass or eye-piece.

Of course the image is nothing like so big as the object. For astronomical objects it is almost infinitely less; still it is an exact representation at an accessible place, and no one expects a telescope to show distant bodies as big as they really are. All it does is to show them bigger than they could be seen without it.

But if the objects are not distant, the same principle may still be applied, and two lenses may be used, one to form an image, the other to magnify it; only if the object can be put where we please, we can easily place it so that its image is already much bigger than the object even before magnification by the eye lens. This is the compound microscope, the invention of which soon followed the telescope. In fact the two instruments shade off into one another, so that the reading telescope or reading microscope of a laboratory (for reading thermometers, and small divisions generally) goes by either name at random.

The arrangement so far described depicts things on the retina the unaccustomed way up. By using a concave glass instead of a convex, and placing it so as to prevent any image being formed, except on the retina direct, this inconvenience is avoided.

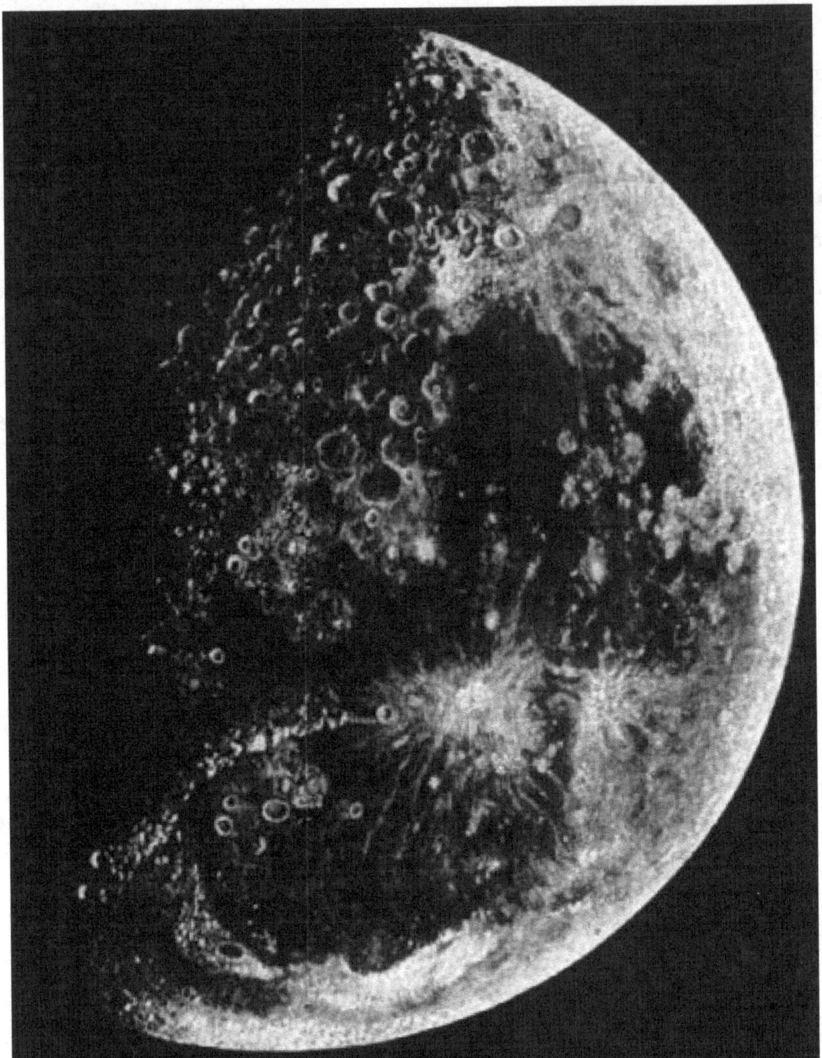

FIG. 38.—View of the half-moon in small telescope. The darker regions, or plains, used to be called "seas."

Such a thing as Galileo made may now be bought at a toy-shop for I suppose half a crown, and yet what a potentiality lay in that "glazed optic tube," as Milton called it. Away he went with it to Venice and showed it to the Signoria, to their great astonishment. "Many noblemen and senators," says Galileo, "though of advanced age, mounted to the top of one of the highest towers to watch the ships, which were visible through my glass two hours before they were seen entering the harbour, for it makes a thing fifty miles off as near and clear as if it were only five." Among the people too the instrument excited the greatest astonishment and interest, so that he was nearly mobbed. The Senate hinted to him that a present of the instrument would not be unacceptable, so Galileo took the hint and made another for them.

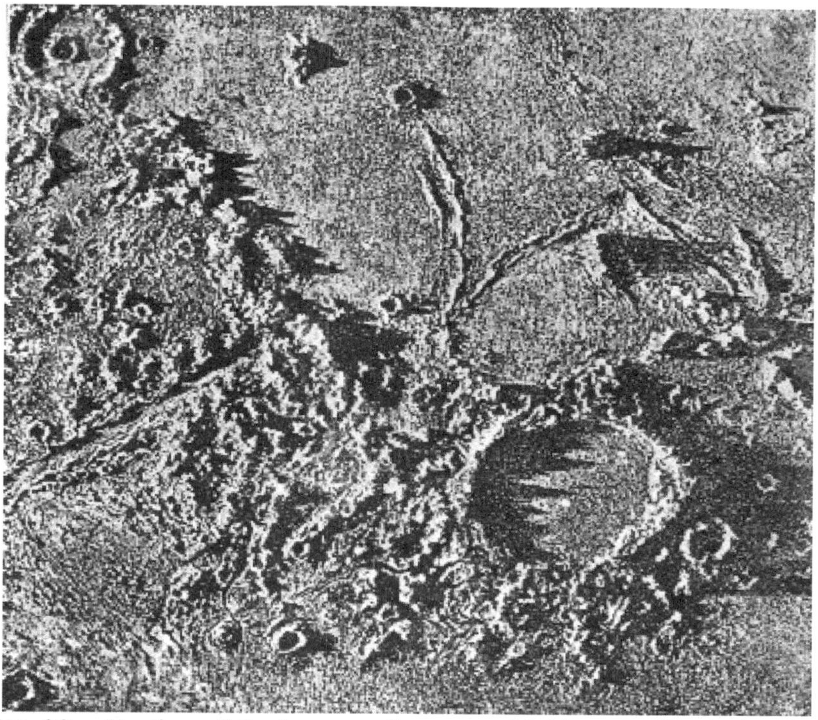

FIG. 39.—Portion of the lunar surface more highly magnified, showing the shadows of a mountain range, deep pits, and other details.

They immediately doubled his salary at Padua, making it 1000 florins, and confirmed him in the enjoyment of it for life.

He now eagerly began the construction of a larger and better instrument. Grinding the lenses with his own hands with consummate skill, he succeeded in making a telescope magnifying thirty times. Thus equipped he was ready to begin a survey of the heavens.

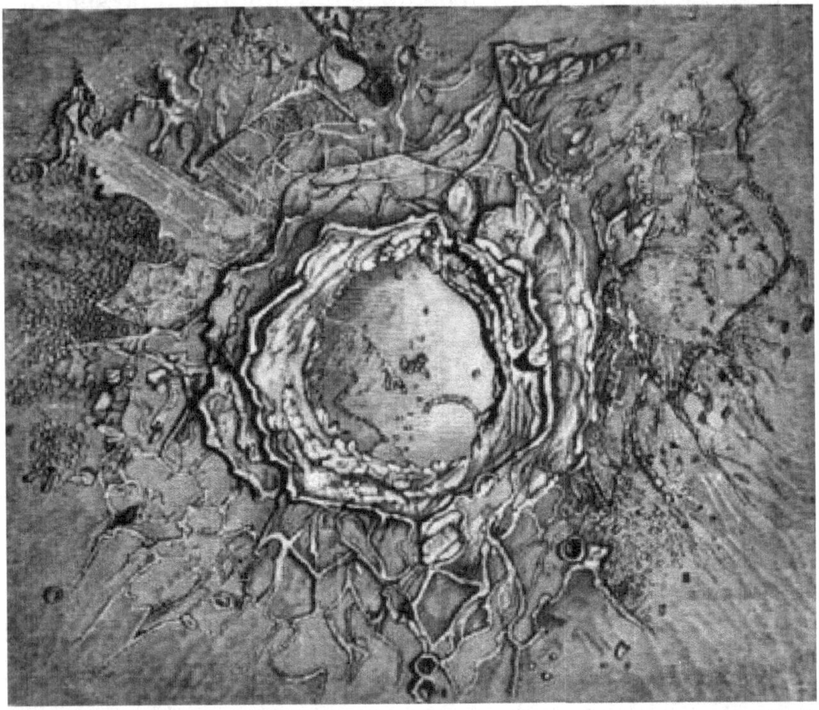

FIG. 40.—Another portion of the lunar surface, showing a so-called crater or vast lava pool and other evidences of ancient heat unmodified by water.

The first object he carefully examined was naturally the moon. He found there everything at first sight very like the earth, mountains and valleys, craters and plains, rocks, and apparently seas. You may imagine the hostility excited among the Aristotelian philosophers, especially no doubt those he had left behind at Pisa, on the ground of his spoiling the pure, smooth, crystalline, celestial face of the moon as they had thought it, and making it harsh and rugged and like so vile and ignoble a body as the earth.

FIG. 41.—Lunar landscape showing earth. The earth would be a stationary object in the moon's sky: its only apparent motion being a slow oscillation as of a pendulum (the result of the moon's libration).

He went further, however, into heterodoxy than this—he not only made the moon like the earth, but he made the earth shine like the moon. The visibility of "the old moon in the new moon's arms" he explained by earth-shine. Leonardo had given the same explanation a century before. Now one of the many stock arguments against Copernican theory of the earth being a planet like the rest was that the earth was dull and dark and did not shine. Galileo argued that it shone just as much as the moon does, and in fact rather more—especially if it be covered with clouds. One reason of the peculiar brilliancy of Venus is that she is a very cloudy planet.[8] Seen from the moon the earth would look exactly as the moon does to us, only a little brighter and sixteen times as big (four times the diameter).

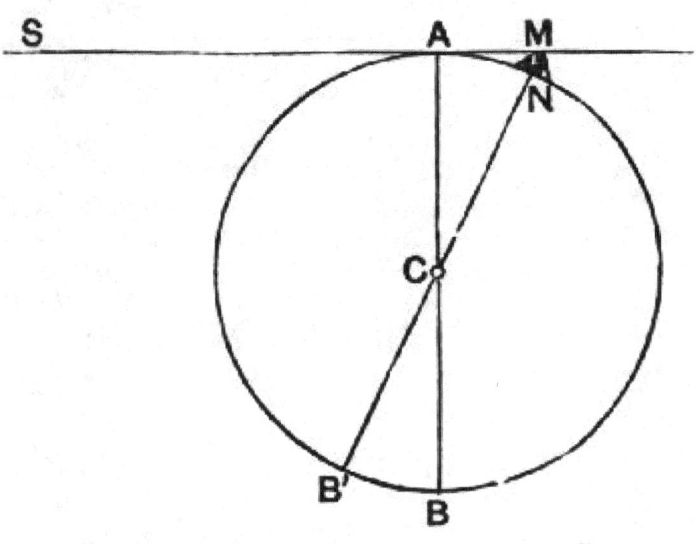

FIG. 42.—Galileo's method of estimating the height of lunar mountain. *AB'BC* is the illuminated half of the moon. *SA* is a solar ray just catching the peak of the mountain *M*. Then by geometry, as *MN* is to *MA*, so is *MA* to *MB'*; whence the height of the mountain, *MN*, can be

determined. The earth and spectator are supposed to be somewhere in the direction *BA* produced, *i.e.* towards the top of the page.

Galileo made a very good estimate of the height of lunar mountains, of which many are five miles high and some as much as seven. He did this simply by measuring from the half-moon's straight edge the distance at which their peaks caught the rising or setting sun. The above simple diagram shows that as this distance is to the diameter of the moon, so is the height of the sun-tipped mountain to the aforesaid distance.

Wherever Galileo turned his telescope new stars appeared. The Milky Way, which had so puzzled the ancients, was found to be composed of stars. Stars that appeared single to the eye were some of them found to be double; and at intervals were found hazy nebulous wisps, some of which seemed to be star clusters, while others seemed only a fleecy cloud.

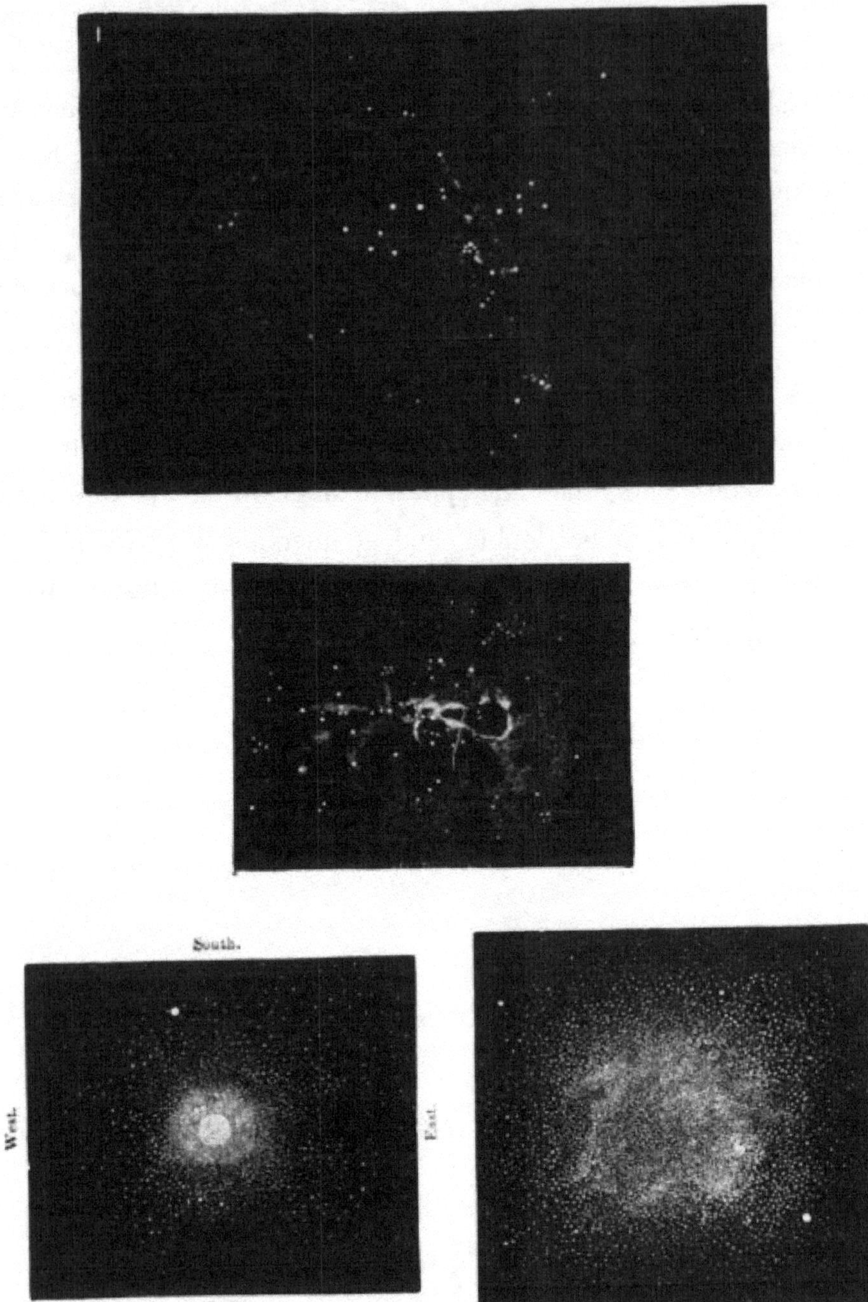

FIG. 43.—Some clusters and nebulæ.

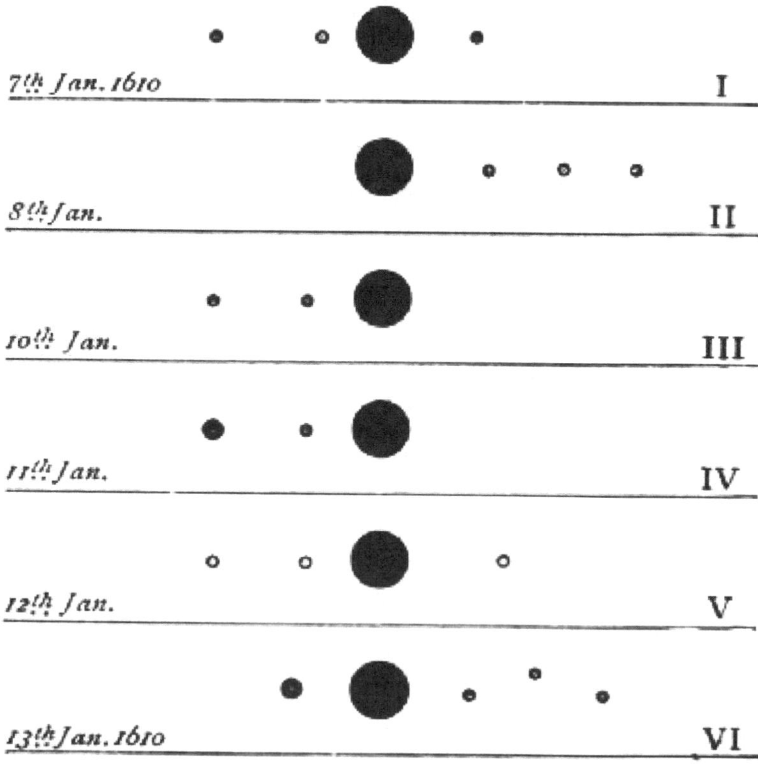

FIG. 44.—Jupiter's satellites, showing the stages of their discovery.

Now we come to his most brilliant, at least his most sensational, discovery. Examining Jupiter minutely on January 7, 1610, he noticed three little stars near it, which he noted down as fixing its then position. On the following night Jupiter had moved to the other side of the three stars. This was natural enough, but was it moving the right way? On examination it appeared not. Was it possible the tables were wrong? The next evening was cloudy, and he had to curb his feverish impatience. On the 10th there were only two, and those on the other side. On the 11th two again, but one bigger than the other. On the 12th the three re-appeared, and on the 13th there were four. No more appeared.

Jupiter then had moons like the earth, four of them in fact, and they revolved round him in periods which were soon determined.

The reason why they were not all visible at first, and why their visibility so rapidly changes, is because they revolve round him almost in the plane of our vision, so that sometimes they are in front and sometimes behind him, while again at other times they plunge into his shadow and are thus eclipsed from the light of the sun which enables us to see them. A large modern telescope will show the moons when in front of Jupiter, but small telescopes will only show them when clear of the disk and shadow. Often all four can be thus seen, but three or two is a very common amount of visibility. Quite a small telescope, such as a ship's telescope, if held steadily, suffices to show the satellites of Jupiter, and very interesting objects they are. They are of habitable size, and may be important worlds for all we know to the contrary.

The news of the discovery soon spread and excited the greatest interest and astonishment. Many of course refused to believe it. Some there were who having been shown them refused to believe their eyes, and asserted that although the telescope acted well enough for terrestrial objects, it was altogether false and illusory when applied to the heavens. Others took the safer ground of refusing to look through the glass. One of these who would not look at the satellites happened to die soon afterwards. "I hope," says Galileo, "that he saw them on his way to heaven."

The way in which Kepler received the news is characteristic, though by adding four to the supposed number of planets it might have seemed to upset his notions about the five regular solids.

He says,[9] "I was sitting idle at home thinking of you, most excellent Galileo, and your letters, when the news was brought me of the discovery of four planets by the help of the double eye-glass. Wachenfels stopped his carriage at my door to tell me, when such a fit of wonder seized me at a report which seemed so very absurd, and I was thrown into such agitation at seeing an old dispute between us decided in this way, that between his joy, my colouring, and the laughter of us both, confounded as we were by such a novelty, we were hardly capable, he of speaking, or I of listening....

"On our separating, I immediately fell to thinking how there could be any addition to the number of planets without overturning my *Mysterium Cosmographicon*, published thirteen years ago, according to which Euclid's five regular solids do not allow more than six planets round the sun.

"But I am so far from disbelieving the existence of the four circumjovial planets that I long for a telescope to anticipate you if possible in discovering two round Mars (as the proportion seems to me to require) six or eight round Saturn, and one each round Mercury and Venus."

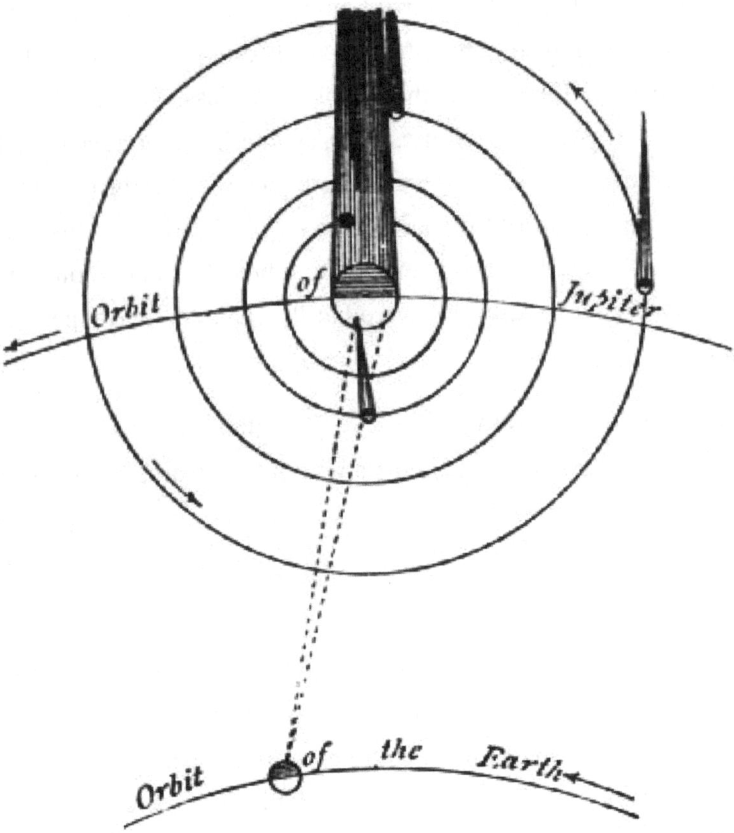

FIG. 45.—Eclipses of Jupiter's satellites. The diagram shows the first (*i.e.* the nearest) moon in Jupiter's shadow, the second as passing between earth and Jupiter, and appearing to transit his disk, the third as on the verge of entering his shadow, and the fourth quite plainly and separately visible.

As an illustration of the opposite school, I will take the following extract from Francesco Sizzi, a Florentine astronomer, who argues against the discovery thus:—

"There are seven windows in the head, two nostrils, two eyes, two ears, and a mouth; so in the heavens there are two favourable stars, two unpropitious, two luminaries, and Mercury alone undecided and indifferent. From which and many other similar phenomena of nature, such as the seven metals, &c., which it were tedious to enumerate, we gather that the number of planets is necessarily seven.

"Moreover, the satellites are invisible to the naked eye, and therefore can have no influence on the earth, and therefore would be useless, and therefore do not exist.

"Besides, the Jews and other ancient nations as well as modern Europeans have adopted the division of the week into seven days, and have named them from the seven planets: now if we increase the number of the planets this whole system falls to the ground."

To these arguments Galileo replied that whatever their force might be as a reason for believing beforehand that no more than seven planets would be discovered, they hardly seemed of sufficient weight to destroy the new ones when actually seen.

Writing to Kepler at this time, Galileo ejaculates:

"Oh, my dear Kepler, how I wish that we could have one hearty laugh together! Here, at Padua, is the principal professor of philosophy whom I have repeatedly and urgently requested to look at the moon and planets through my glass, which he pertinaciously refuses to do. Why are you not here? What shouts of laughter we should have at this glorious folly! And to hear the professor of philosophy at Pisa labouring before the grand duke with logical arguments, as if with magical incantations, to charm the new planets out of the sky."

A young German *protégé* of Kepler, Martin Horkey, was travelling in Italy, and meeting Galileo at Bologna was favoured with a view through his telescope. But supposing that Kepler must necessarily be jealous of such great discoveries, and thinking to please him, he writes, "I cannot tell what to think about these observations. They are stupendous, they are wonderful, but whether they are true or false I cannot tell." He concludes, "I will never concede his four new planets to that Italian from Padua though I die for it." So he published a pamphlet asserting that reflected rays and optical illusions were the sole cause of the appearance, and that the only use of the imaginary planets was to gratify Galileo's thirst for gold and notoriety.

When after this performance he paid a visit to his old instructor Kepler, he got a reception which astonished him. However, he pleaded so hard to be forgiven that Kepler restored him to partial favour, on this condition, that he was to look again at the satellites, and this time to see them and own that they were there.

By degrees the enemies of Galileo were compelled to confess to the truth of the discovery, and the next step was to outdo him. Scheiner counted five, Rheiter nine, and others went as high as twelve. Some of these were imaginary, some were fixed stars, and four satellites only are known to this day.[10]

Here, close to the summit of his greatness, we must leave him for a time. A few steps more and he will be on the brow of the hill; a short piece of table-land, and then the descent begins.

LECTURE V

GALILEO AND THE INQUISITION

ONE sinister event occurred while Galileo was at Padua, some time before the era we have now arrived at, before the invention of the telescope—two years indeed after he had first gone to Padua; an event not directly concerning Galileo, but which I must mention because it must have shadowed his life both at the time and long afterwards. It was the execution of Giordano Bruno for heresy. This eminent philosopher had travelled largely, had lived some time in England, had acquired new and heterodox views on a variety of subjects, and did not hesitate to propound them even after he had returned to Italy.

The Copernican doctrine of the motion of the earth was one of his obnoxious heresies. Being persecuted to some extent by the Church, Bruno took refuge in Venice—a free republic almost independent of the Papacy—where he felt himself safe. Galileo was at Padua hard by: the University of Padua was under the government of the Senate of Venice: the two men must in all probability have met.

Well, the Inquisition at Rome sent messengers to Venice with a demand for the extradition of Bruno—they wanted him at Rome to try him for heresy.

In a moment of miserable weakness the Venetian republic gave him up, and Bruno was taken to Rome. There he was tried, and cast into the dungeons for six years, and because he entirely refused to recant, was at length delivered over to the secular arm and burned at the stake on 16th February, Anno Domini 1600.

This event could not but have cast a gloom over the mind of lovers and expounders of truth, and the lesson probably sank deep into Galileo's soul.

In dealing with these historic events will you allow me to repudiate once for all the slightest sectarian bias or meaning. I have nothing to do with Catholic or Protestant as such. I have nothing to do with the Church of Rome as such. I am dealing with the history of science. But historically at one period science and the Church came into conflict. It was not specially one Church rather than another—it was the Church in general, the only one that then existed in those countries. Historically, I say, they came into conflict, and historically the Church was the conqueror. It got its way; and science, in the persons of Bruno, Galileo, and several others, was vanquished.

Such being the facts, there is no help but to mention them in dealing with the history of science.

Doubtless *now* the Church regards it as an unhappy victory, and gladly would ignore this painful struggle. This, however, is impossible. With their creed the Churchmen of that day could act in no other way. They were bound to prosecute heresy, and they were bound to conquer in the struggle or be themselves shattered.

But let me insist on the fact that no one accuses the ecclesiastical courts of crime or evil motives. They attacked heresy after their manner, as the civil courts attacked witchcraft after *their* manner. Both erred grievously, but both acted with the best intentions.

We must remember, moreover, that his doctrines were scientifically heterodox, and the University Professors of that day were probably quite as ready to condemn them as the Church was. To realise the position we must think of some subjects which *to-day* are scientifically heterodox, and of the customary attitude adopted towards them by persons of widely differing creeds.

If it be contended now, as it is, that the ecclesiastics treated Galileo well, I admit it freely: they treated him as well as they possibly could. They overcame him, and he recanted; but if he had not recanted, if he had persisted in his heresy, they would—well, they would still have treated his soul well, but they would have set fire to his body. Their mistake consisted not in cruelty, but in supposing themselves the arbiters of eternal truth; and by no amount of slurring and glossing over facts can they evade the responsibility assumed by them on account of this mistaken attitude.

I am not here attacking the dogma of Papal Infallibility: it is historically, I believe, quite unaffected by the controversy respecting the motion of the earth, no Papal edict *ex cathedrâ* having been promulgated on the subject.

We left Galileo standing at his telescope and beginning his survey of the heavens. We followed him indeed through a few of his first great discoveries—the discovery of the mountains and other variety of surface in the moon, of the nebulæ and a multitude of faint stars, and lastly of the four satellites of Jupiter.

This latter discovery made an immense sensation, and contributed its share to his removal from Padua, which quickly followed it, as I shall shortly narrate; but first I think it will be best to continue our survey of his astronomical discoveries without regard to the place whence they were made.

Before the end of the year Galileo had made another discovery—this time on Saturn. But to guard against the host of plagiarists and impostors, he published it in the form of an anagram, which, at the request of the Emperor Rudolph (a request probably inspired by Kepler), he interpreted; it ran thus: The furthest planet is triple.

Very soon after he found that Venus was changing from a full moon to a half moon appearance. He announced this also by an anagram, and waited till it should become a crescent, which it did.

This was a dreadful blow to the anti-Copernicans, for it removed the last lingering difficulty to the reception of the Copernican doctrine.

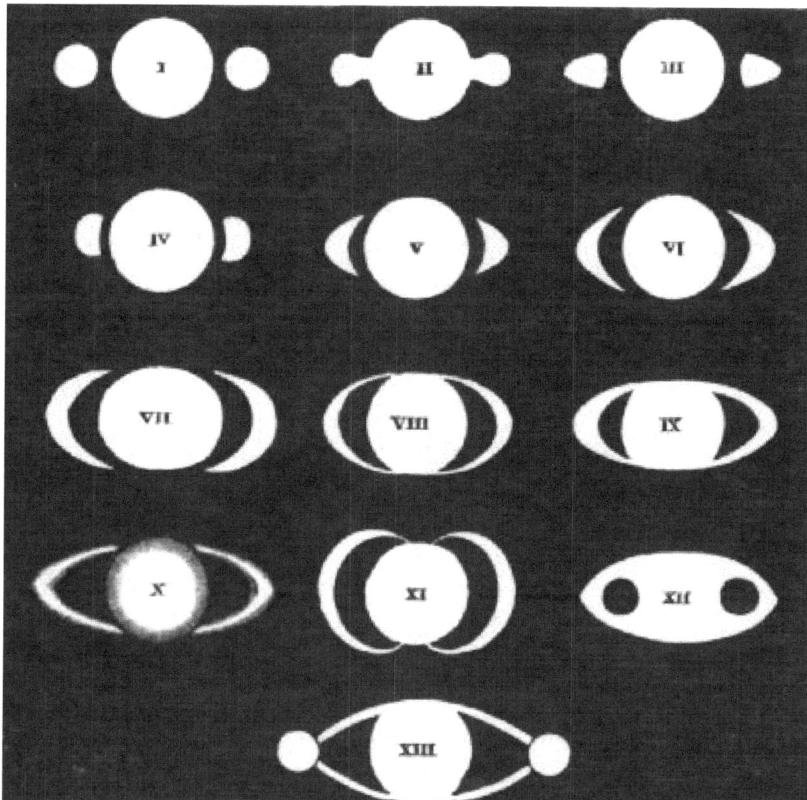

FIG. 46.—Old drawings of Saturn by different observers, with the imperfect instruments of that day. The first is Galileo's idea of what he saw.

Copernicus had predicted, indeed, a hundred years before, that, if ever our powers of sight were sufficiently enhanced, Venus and Mercury would be seen to have phases like the moon. And now Galileo with his telescope verifies the prediction to the letter.

Here was a triumph for the grand old monk, and a bitter morsel for his opponents.

Castelli writes: "This must now convince the most obstinate." But Galileo, with more experience, replies:—"You almost make me laugh by saying that these clear observations are sufficient to convince the most obstinate; it seems you have yet to learn that long ago the observations were enough to convince those who are capable of reasoning, and those who wish to learn the truth; but that to convince the obstinate, and those who care for nothing beyond the vain applause of the senseless vulgar, not even the testimony of the stars would suffice, were they to descend on earth to speak for themselves. Let us, then, endeavour to procure some knowledge for ourselves, and rest contented with this sole satisfaction; but of advancing in popular opinion, or of gaining the assent of the book-philosophers, let us abandon both the hope and the desire."

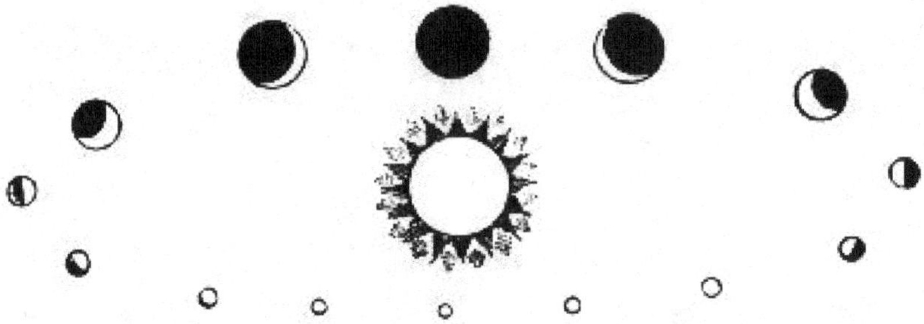

FIG. 47.—Phases of Venus. Showing also its apparent variations in size by reason of its varying distance from the earth. When fully illuminated it is necessarily most distant. It looks brightest to us when a broad crescent.

What a year's work it had been!

In twelve months observational astronomy had made such a bound as it has never made before or since.

Why did not others make any of these observations? Because no one could make telescopes like Galileo.

He gathered pupils round him however, and taught them how to work the lenses, so that gradually these instruments penetrated Europe, and astronomers everywhere verified his splendid discoveries.

But still he worked on, and by March in the very next year, he saw something still more hateful to the Aristotelian philosophers, viz. spots on the sun.

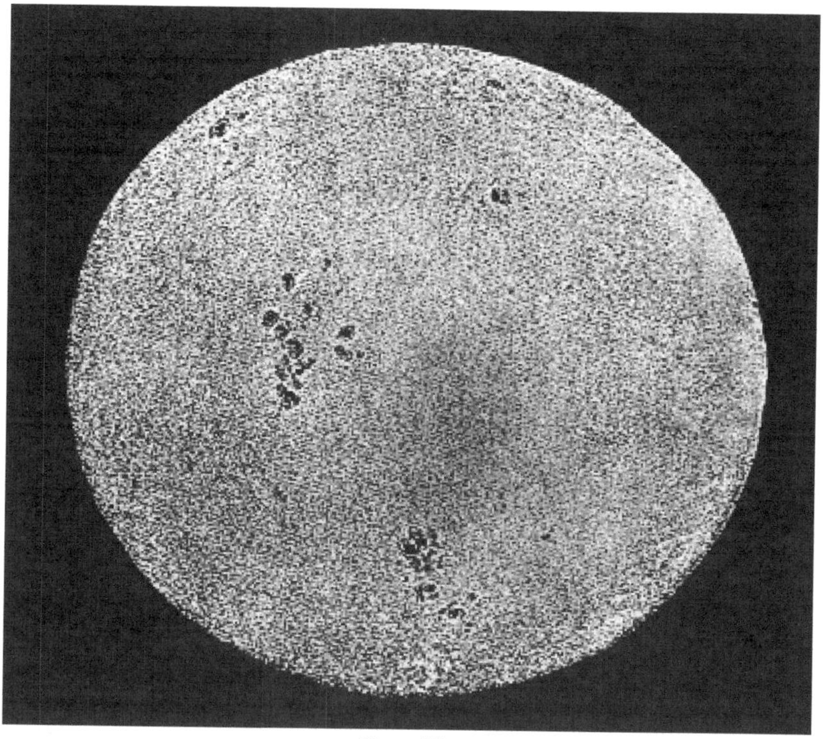

FIG. 48.

If anything was pure and perfect it was the sun, they said. Was this impostor going to blacken its face too?

Well, there they were. They slowly formed and changed, and by moving all together showed him that the sun rotated about once a month.

Before taking leave of Galileo's astronomical researches, I must mention an observation made at the end of 1612, that the apparent triplicity of Saturn (Fig. 46) had vanished.

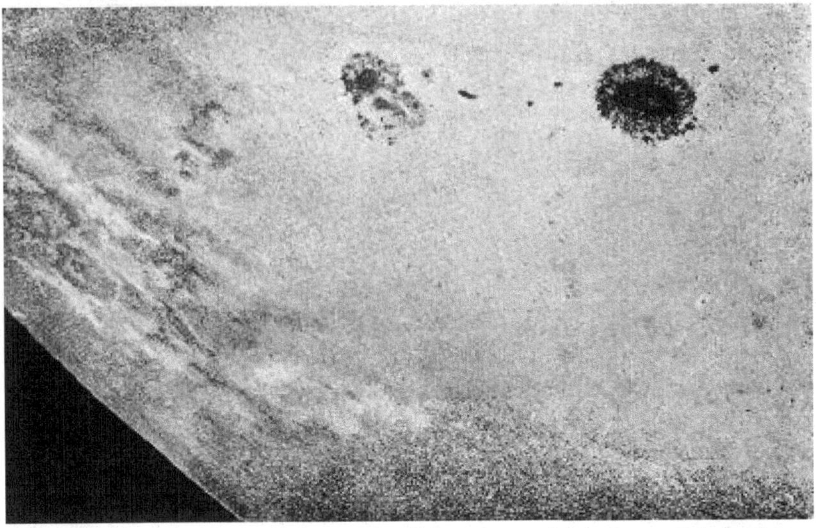

FIG. 49.—A portion of the sun's disk as seen in a powerful modern telescope.

"Looking on Saturn within these few days, I found it solitary, without the assistance of its accustomed stars, and in short perfectly round and defined, like Jupiter, and such it still remains. Now what can be said of so strange a metamorphosis? Are perhaps the two smaller stars consumed like spots on the sun? Have they suddenly vanished and fled? Or has Saturn devoured his own children? Or was the appearance indeed fraud and illusion, with which the glasses have so long time mocked me and so many others who have so often observed with me? Now perhaps the time is come to revive the withering hopes of those, who, guided by more profound contemplations, have fathomed all the fallacies of the new observations and recognized their impossibility! I cannot resolve what to say in a chance so strange, so new, so unexpected. The shortness of time, the unexampled occurrence, the weakness of my intellect, the terror of being mistaken, have greatly confounded me."

However, he plucked up courage, and conjectured that the two attendants would reappear, by revolving round the planet.

FIG. 50.—Saturn and his rings, as seen under the most favourable circumstances.

The real reason of their disappearance is well known to us now. The plane of Saturn's rings oscillates slowly about our line of sight, and so we sometimes see them edgeways and sometimes with a moderate amount of obliquity. The rings are so thin that, when turned precisely edgeways, they become invisible. The two imaginary attendants were the most conspicuous portions of the ring, subsequently called *ansæ*.

I have thought it better not to interrupt this catalogue of brilliant discoveries by any biographical details; but we must now retrace our steps to the years 1609 and 1610, the era of the invention of the telescope.

By this time Galileo had been eighteen years at Padua, and like many another man in like case, was getting rather tired of continual lecturing. Moreover, he felt so full of ideas that he longed to have a better opportunity of following them up, and more time for thinking them out.

Now in the holidays he had been accustomed to return to his family home at Pisa, and there to come a good deal into contact with the Grand-Ducal House of Tuscany. Young Cosmo di Medici became in fact his pupil, and arrived at man's estate with the highest opinion of the philosopher. This young man had now come to the throne as Cosmo II., and to him Galileo wrote saying how much he should like more time and leisure, how full he was of discoveries if he only had the chance of a reasonable income without the necessity of consuming so large a portion of his time in elementary teaching, and practically asking to be removed to some position in the Court. Nothing was done for a time, but negotiations proceeded, and soon after the discovery of Jupiter's satellites Cosmo wrote making a generous offer, which Galileo gladly and enthusiastically accepted, and at once left Padua for Florence. All his subsequent discoveries date from Florence.

Thus closed his brilliant and happy career as a professor at the University of Padua. He had been treated well: his pay had become larger than that of any Professor of Mathematics up to that time; and, as you know, immediately after his invention of the telescope the Venetian Senate, in a fit of enthusiasm, had doubled it and secured it to him for life wherever he was. To throw up his chair and leave the place the very next year scarcely seems a strictly honourable procedure. It was legal enough no doubt, and it is easy for small men to criticize a great one, but nevertheless I think we must admit that it is a step such as a man with a keen sense of honour would hardly have taken.

One quite feels and sympathizes with the temptation. Not emolument, but leisure; freedom from harassing engagements and constant teaching, and liberty to prosecute his studies day and night without interference: this was the golden prospect before him. He yielded, but one cannot help wishing he had not.

As it turned out it was a false step—the first false step of his public career. When made it was irretrievable, and it led to great misery.

At first it seemed brilliant enough. The great philosopher of the Tuscan Court was courted and flattered by princes and nobles, he enjoyed a world-wide reputation, lived as luxuriously as he cared for, had his time all to himself, and lectured but very seldom, on great occasions or to a few crowned heads.

His position was in fact analogous to that of Tycho Brahé in his island of Huen.

Misfortune overtook both. In Tycho's case it arose mainly from the death of his patron. In Galileo's it was due to a more insidious cause, to understand which cause aright we must remember the political divisions of Italy at that date.

Tuscany was a Papal State, and thought there was by no means free. Venice was a free republic, and was even hostile to the Papacy. In 1606 the Pope had placed it under an interdict. In reply it had ejected every Jesuit.

Out of this atmosphere of comparative enlightenment and freedom into that hotbed of mediævalism and superstition went Galileo with his eyes open. Keen was the regret of his Paduan and Venetian friends; bitter were their remonstrances and exhortations. But he was determined to go, and, not without turning some of his old friends into enemies, he went.

Seldom has such a man made so great a mistake: never, I suppose, has one been so cruelly punished for it.

FIG. 51.—Map of Italy.

We must remember, however, that Galileo, though by no means a saint, was yet a really religious man, a devout Catholic and thorough adherent of the Church, so that he would have no dislike to place himself under her sway. Moreover, he had been born a Tuscan, his family had lived at Florence or Pisa, and it felt like going home. His theological attitude is worthy of notice, for he was not in the least a sceptic. He quite acquiesces in the authority of the Bible, especially in all matters concerning faith and conduct; as to its statements in scientific matters, he argues that we are so liable to misinterpret their meaning that it is really easier to examine Nature for truth in scientific matters, and that when direct observation and

Scripture seem to clash, it is because of our fallacious interpretation of one or both of them. He is, in fact, what one now calls a "reconciler."

It is curious to find such a man prosecuted for heresy, when to-day his opinions are those of the orthodox among the orthodox. But so it ever is, and the heresy of one generation becomes the commonplace of the next.

He accepts Joshua's miracle, for instance, not as a striking poem, but as a literal fact; and he points out how much more simply it could be done on the Copernican system by stopping the earth's rotation for a short time, than by stopping the sun and moon and all the host of heaven as on the old Ptolemaic system, or again by stopping only the sun and not any of the other bodies, and so throwing astronomy all wrong.

This reads to us like satire, but no doubt it was his genuine opinion.

These Scriptural reconciliations of his, however, angered the religious authorities still more. They said it was bad enough for this heretic to try and upset old *scientific* beliefs, and to spoil the face of *Nature* with his infidel discoveries, but at least he might leave the Bible alone; and they addressed an indignant remonstrance to Rome, to protect it from the hands of ignorant laymen.

Thus, wherever he turned he encountered hostility. Of course he had many friends—some of them powerful like Cosmo, all of them faithful and sincere. But against the power of Rome what could they do? Cosmo dared no more than remonstrate, and ultimately his successor had to refrain from even this, so enchained and bound was the spirit of the rulers of those days; and so when his day of tribulation came he stood alone and helpless in the midst of his enemies.

You may wonder, perhaps, why this man should excite so much more hostility than many another man who was suffered to believe and teach much the same doctrines unmolested. But no other man had made such brilliant and exciting discoveries. No man stood so prominently forward in the eyes of all Christendom as the champion of the new doctrines. No other man stated them so clearly and forcibly, nor drove them home with such brilliant and telling illustrations.

And again, there was the memory of his early conflict with the Aristotelians at Pisa, of his scornful and successful refutation of their absurdities. All this made him specially obnoxious to the Aristotelian Jesuits in their double capacity both of priests and of philosophers, and they singled him out for relentless official persecution.

Not yet, however, is he much troubled by them. The chief men at Rome have not yet moved. Messages, however, keep going up from Tuscany to Rome respecting the teachings of this man, and of the harm he is doing by his pertinacious preaching of the Copernican doctrine that the earth moves.

At length, in 1615, Pope Paul V. wrote requesting him to come to Rome to explain his views. He went, was well received, made a special friend of Cardinal Barberino—an accomplished man in high position, who became in fact the next Pope. Galileo showed cardinals and others his telescope, and to as many as would look through it he showed Jupiter's satellites and his other discoveries. He had a most successful visit. He talked, he harangued, he held forth in the midst of fifteen or twenty disputants at once, confounding his opponents and putting them to shame.

His method was to let the opposite arguments be stated as fully and completely as possible, himself aiding, and often adducing the most forcible and plausible arguments against his own views; and then, all having been well stated, he would proceed to utterly undermine and demolish the whole fabric, and bring out the truth in such a way as to convince all honest minds. It was this habit that made him such a formidable antagonist. He never shrank from meeting an opposing argument, never sought to ignore it, or cloak it in a cloud of words. Every hostile argument he seemed to delight in, as a foe to be crushed, and the better and stronger they sounded the more he liked them. He knew many of them well, he invented a number more, and had he chosen could have out-argued the stoutest Aristotelian on his own grounds. Thus did he lead his adversaries on, almost like Socrates, only to ultimately overwhelm them in a more hopeless rout. All this in Rome too, in the heart of the Catholic world. Had he been worldly-wise, he would certainly have kept silent and unobtrusive till he had leave to go away again. But he felt like an apostle of the new doctrines, whose mission it was to proclaim them even in this centre of the world and of the Church.

Well, he had an audience with the Pope—a chat an hour long—and the two parted good friends, mutually pleased with each other.

He writes that he is all right now, and might return home when he liked. But the question began to be agitated whether the whole system of Copernicus ought not to be condemned as impious and heretical. This view was persistently urged upon the Pope and College of Cardinals, and it was soon to be decided upon.

Had Galileo been unfaithful to the Church he could have left them to stultify themselves in any way they thought proper, and himself have gone; but he felt supremely interested in the result, and he stayed. He writes:—

"So far as concerns the clearing of my own character, I might return home immediately; but although this new question regards me no more than all those who for the last eighty years have supported those opinions both in public and private, yet, as perhaps I may be of some assistance in that part of the discussion which depends on the knowledge of truths ascertained by means of the sciences which I profess, I, as a zealous and Catholic Christian, neither can nor ought to withhold that assistance which my knowledge affords, and this business keeps me sufficiently employed."

It is possible that his stay was the worst thing for the cause he had at heart. Anyhow, the result was that the system was condemned, and both the book of Copernicus and the epitome of it by Kepler were placed on the forbidden list,[11] and Galileo himself was formally ordered never to teach or to believe the motion of the earth.

He quitted Rome in disgust, which before long broke out in satire. The only way in which he could safely speak of these views now was as if they were hypothetical and uncertain, and so we find him writing to the Archduke Leopold, with a presentation copy of his book on the tides, the following:—

"This theory occurred to me when in Rome whilst the theologians were debating on the prohibition of Copernicus's book, and of the opinion maintained in it of the motion of the earth, which I at that time believed: until it pleased those gentlemen to suspend the book, and declare the opinion false and repugnant to the Holy Scriptures. Now, as I know how well it becomes me to obey and believe the decisions of my superiors, which proceed out of more knowledge than the weakness of my intellect can attain to, this theory which I send you, which is founded on the motion of the earth, I now look upon as a fiction and a dream, and beg your highness to receive it as such. But as poets often learn to prize the creations of their fancy, so in like manner do I set some value on this absurdity of mine. It is true that when I sketched this little work I did hope that Copernicus would not, after eighty years, be convicted of error; and I had intended to develop and amplify it further, but a voice from heaven suddenly awakened me, and at once annihilated all my confused and entangled fancies."

This sarcasm, if it had been in print, would probably have been dangerous. It was safe in a private letter, but it shows us his real feelings.

However, he was left comparatively quiet for a time. He was getting an old man now, and passed the time studiously enough, partly at his house in Florence, partly at his villa in Arcetri, a mile or so out of the town.

Here was a convent, and in it his two daughters were nuns. One of them, who passed under the name of Sister Maria Celeste, seems to have been a woman of considerable capacity—certainly she was of a most affectionate disposition—and loved and honoured her father in the most dutiful way.

This was a quiet period of his life, spoiled only by occasional fits of illness and severe rheumatic pains, to which the old man was always liable. Many little circumstances are known of this peaceful time. For instance, the convent clock won't go, and Galileo mends it for them. He is always doing little things for them, and sending presents to the Lady Superior and his two daughters.

He was occupied now with problems in hydrostatics, and on other matters unconnected with astronomy: a large piece of work which I must pass over. Most interesting and acute it is, however.

In 1623, when the old Pope died, there was elected to the Papal throne, as Urban VIII., Cardinal Barberino, a man of very considerable enlightenment, and a personal friend of Galileo's, so that both he and his daughters rejoice greatly, and hope that things will come all right, and the forbidding edict be withdrawn.

The year after this election he manages to make another journey to Rome to compliment his friend on his elevation to the Pontifical chair. He had many talks with Urban, and made himself very agreeable.

Urban wrote to the Grand Duke Ferdinand, son of Cosmo:—

"For We find in him not only literary distinction but also love of piety, and he is strong in those qualities by which Pontifical good will is easily obtainable. And now, when he has been brought to this city to congratulate Us on Our elevation, We have very lovingly embraced him; nor can We suffer him to return to the country whither your liberality recalls him without an ample provision of Pontifical love. And that you may know how dear he is to Us, We have willed to give him this honourable testimonial of virtue and piety. And We further signify that every benefit which you shall confer upon him, imitating or even surpassing your father's liberality, will conduce to Our gratification."

Encouraged, doubtless, by these marks of approbation, and reposing too much confidence in the individual good will of the Pope, without heeding the crowd of half-declared enemies who were seeking to undermine his reputation, he set about, after his return to Florence, his greatest literary and most popular work, *Dialogues on the Ptolemaic and Copernican Systems*. This purports to be a series of four conversations between three characters: Salviati, a Copernican philosopher; Sagredo, a wit and scholar, not specially learned, but keen and critical, and who lightens the talk with chaff; Simplicio, an Aristotelian philosopher, who propounds the stock absurdities which served instead of arguments to the majority of men.

The conversations are something between Plato's *Dialogues* and Sir Arthur Helps's *Friends in Council*. The whole is conducted with great good temper and fairness; and, discreetly enough, no definite conclusion is arrived at, the whole being left in abeyance as if for a fifth and decisive dialogue, which, however, was never written, and perhaps was only intended in case the reception was favourable.

The preface also sets forth that the object of the writer is to show that the Roman edict forbidding the Copernican doctrine was not issued in ignorance of the facts of the case, as had been maliciously reported, and that he wishes to show how well and clearly it was all known beforehand. So he says the dialogue on the Copernican side takes up the question purely as a mathematical hypothesis or speculative figment, and gives it every artificial advantage of which the theory is capable.

This piece of caution was insufficient to blind the eyes of the Cardinals; for in it the arguments in favour of the earth's motion are so cogent and unanswerable, and are so popularly stated, as to do more in a few years to undermine the old system than all that he had written and spoken before. He could not get it printed for two years after he had written it, and then only got consent through a piece of carelessness or laziness on the part of the ecclesiastical censor through whose hands the manuscript passed—for which he was afterwards dismissed.

However, it did appear, and was eagerly read; the more, perhaps, as the Church at once sought to suppress it.

The Aristotelians were furious, and represented to the Pope that he himself was the character intended by Simplicio, the philosopher whose opinions get alternately refuted and ridiculed by the other two, till he is reduced to an abject state of impotence.

The idea that Galileo had thus cast ridicule upon his friend and patron is no doubt a gratuitous and insulting libel: there is no telling whether or not Urban believed it, but certainly his countenance changed to Galileo henceforward, and whether overruled by his Cardinals, or actuated by some other motive, his favour was completely withdrawn.

The infirm old man was instantly summoned to Rome. His friends pleaded his age—he was now seventy—his ill-health, the time of year, the state of the roads, the quarantine existing on account of the plague. It was all of no avail, to Rome he must go, and on the 14th of February he arrived.

FIG. 52.—Portrait of Galileo.

His daughter at Arcetri was in despair; and anxiety and fastings and penances self-inflicted on his account, dangerously reduced her health.

At Rome he was not imprisoned, but he was told to keep indoors, and show himself as little as possible. He was allowed, however, to stay at the house of the Tuscan Ambassador instead of in gaol.

By April he was removed to the chambers of the Inquisition, and examined several times. Here, however, the anxiety was too much, and his health began to give way seriously; so, before long, he was allowed to return to the Ambassador's house; and, after application had been made, was allowed to drive in the public garden in a half-closed carriage. Thus in every way the Inquisition dealt with him as leniently as they could. He was now their prisoner, and they might have cast him into their dungeons, as many another had been cast. By whatever they were influenced—perhaps the Pope's old friendship, perhaps his advanced age and infirmities—he was not so cruelly used.

Still, they had their rules; he *must* be made to recant and abjure his heresy; and, if necessary, torture must be applied. This he knew well enough, and his daughter knew it, and her distress may be imagined. Moreover, it is not as if they had really been heretics, as if they hated or despised the Church of Rome. On the contrary, they loved and honoured the Church. They were sincere and devout worshippers, and only on a few scientific matters did Galileo presume to differ from his ecclesiastical superiors: his disagreement with them occasioned him real sorrow; and his dearest hope was that they could be brought to his way of thinking and embrace the truth.

Every time he was sent for by the Inquisition he was in danger of torture unless he recanted. All his friends urged him repeatedly to submit. They said resistance was hopeless and fatal. Within the memory of men still young, Giordano Bruno had been burnt alive for a similar heresy. This had happened while Galileo was at Padua. Venice was full of it. And since that, only eight years ago indeed, Antonio de Dominis, Archbishop of Salpetria, had been sentenced to the same fate: "to be handed over to the secular arm to be dealt with as mercifully as possible without the shedding of blood." So ran the hideous formula condemning a man to the stake. After his sentence, this unfortunate man died in the dungeons in which he had been incarcerated six years—died what is called a "natural" death; but the sentence was carried out, notwithstanding, on his lifeless body and his writings. His writings for which he had been willing to die!

These were the tender mercies of the Inquisition; and this was the kind of meaning lurking behind many of their well-sounding and merciful phrases. For instance, what they call "rigorous examination," we call "torture." Let us, however, remember in our horror at this mode of compelling a prisoner to say anything they wished, that they were a legally constituted tribunal; that they acted with well established rules, and not in passion; and that torture was a recognized mode of extracting evidence, not only in ecclesiastical but in civil courts, at that date.

All this, however, was but poor solace to the pitiable old philosopher, thus ruthlessly haled up and down, questioned and threatened, threatened and questioned, receiving agonizing letters from his daughter week by week, and trying to keep up a little spirit to reply as happily and hopefully as he could.

This condition of things could not go on. From February to June the suspense lasted. On the 20th of June he was summoned again, and told he would be wanted all next day for a rigorous examination. Early in the morning of the 21st he repaired thither, and the doors were shut. Out of those chambers of horror he did not reappear till the 24th. What went on all those three days no one knows. He himself was bound to secrecy. No outsider was present. The records of the Inquisition are jealously guarded. That he was technically tortured is certain; that he actually underwent the torment of the rack is doubtful. Much learning has been expended upon the question, especially in Germany. Several eminent scholars have held the fact of actual torture to be indisputable (geometrically certain, one says), and they confirm it by the hernia from which he afterwards suffered, this being a well-known and frequent consequence.

Other equally learned commentators, however, deny that the last stage was reached. For there are five stages all laid down in the rules of the Inquisition, and steadily adhered to in a rigorous examination, at each stage an opportunity being given for recantation, every utterance, groan, or sigh being strictly recorded. The recantation so given has to be confirmed a day or two later, under pain of a precisely similar ordeal.

The five stages are:—1st. The official threat in the court. 2nd. The taking to the door of the torture chamber and renewing the official threat. 3rd. The taking inside and showing the instruments. 4th. Undressing and binding upon the rack. 5th. *Territio realis.*

Through how many of these ghastly acts Galileo passed I do not know. I hope and believe not the last.

There are those who lament that he did not hold out, and accept the crown of martyrdom thus offered to him. Had he done so we know his fate—a few years' languishing in the dungeons, and then the flames.

Whatever he ought to have done, he did not hold out—he gave way. At one stage or another of the dread ordeal he said: "I am in your hands. I will say whatever you wish." Then was he removed to a cell while his special form of perjury was drawn up.

The next day, clothed as a penitent, the venerable old man was taken to the Convent of Minerva, where the Cardinals and prelates were assembled for the purpose of passing judgment upon him.

The text of the judgment I have here, but it is too long to read. It sentences him—1st. To the abjuration. 2nd. To formal imprisonment for life. 3rd. To recite the seven penitential psalms every week.

Ten Cardinals were present; but, to their honour be it said, three refused to sign; and this blasphemous record of intolerance and bigoted folly goes down the ages with the names of seven Cardinals immortalized upon it.

This having been read, he next had to read word for word the abjuration which had been drawn up for him, and then sign it.

The Abjuration of Galileo.

"I, Galileo Galilei, son of the late Vincenzo Galilei, of Florence, aged seventy years, being brought personally to judgment, and kneeling before you Most Eminent and Most Reverend Lords Cardinals, General Inquisitors of the universal Christian republic against heretical depravity, having before my eyes the Holy Gospels, which I touch with my own hands, swear that I have always believed, and now believe, and with the help of God will in future believe, every article which the Holy Catholic and Apostolic Church of Rome holds, teaches, and preaches. But because I have been enjoined by this Holy Office altogether to abandon the false opinion which maintains that the sun is the centre and immovable, and forbidden to hold, defend, or teach the said false doctrine in any manner, and after it hath been signified to me that the said doctrine is repugnant with the Holy Scripture, I have written and printed a book, in which I treat of the same doctrine now condemned, and adduce reasons with great force in support of the same, without giving any solution, and therefore have been judged grievously suspected of heresy; that is to say, that I held and believed that the sun is the centre of the universe and is immovable, and that the earth is not the centre and is movable; willing, therefore, to remove from the minds of your Eminences, and of every Catholic Christian, this vehement suspicion rightfully entertained towards me, with a sincere heart and unfeigned faith, I abjure, curse, and detest the said errors and heresies, and generally every other error and sect contrary to Holy Church; and I swear that I will never more in future say or assert anything verbally, or in writing, which may give rise to a similar suspicion of me; but if I shall know any heretic, or any one suspected of heresy, that I will denounce him to this Holy Office, or to the Inquisitor or Ordinary of the place where I may be; I swear, moreover, and promise, that I will fulfil and observe fully, all the penances which have been or shall be laid on me by this Holy Office. But if it shall happen that I violate any of my said promises, oaths, and protestations (which God avert!), I subject

myself to all the pains and punishments which have been decreed and promulgated by the sacred canons, and other general and particular constitutions, against delinquents of this description. So may God help me, and his Holy Gospels which I touch with my own hands. I, the above-named Galileo Galilei, have abjured, sworn, promised, and bound myself as above, and in witness thereof with my own hand have subscribed this present writing of my abjuration, which I have recited word for word. At Rome, in the Convent of Minerva, 22nd June, 1633. I, Galileo Galilei, have abjured as above with my own hand."

Those who believe the story about his muttering to a friend, as he rose from his knees, "e pur si muove," do not realize the scene.

1st. There was no friend in the place.

2nd. It would have been fatally dangerous to mutter anything before such an assemblage.

3rd. He was by this time an utterly broken and disgraced old man; wishful, of all things, to get away and hide himself and his miseries from the public gaze; probably with his senses deadened and stupefied by the mental sufferings he had undergone, and no longer able to think or care about anything—except perhaps his daughter,—certainly not about any motion of this wretched earth.

Far and wide the news of the recantation spread. Copies of the abjuration were immediately sent to all Universities, with instructions to the professors to read it publicly.

At Florence, his home, it was read out in the Cathedral church, all his friends and adherents being specially summoned to hear it.

For a short time more he was imprisoned in Rome; but at length was permitted to depart, never more of his own will to return.

He was allowed to go to Siena. Here his daughter wrote consolingly, rejoicing at his escape, and saying how joyfully she already recited the penitential psalms for him, and so relieved him of that part of his sentence.

But the poor girl was herself, by this time, ill—thoroughly worn out with anxiety and terror; she lay, in fact, on what proved to be her death-bed. Her one wish was to see her dearest lord and father, so she calls him, once more. The wish was granted. His prison was changed, by orders from Rome, from Siena to Arcetri, and once more father and daughter embraced. Six days after this she died.

The broken-hearted old man now asks for permission to go to live in Florence, but is met with the stern answer that he is to stay at Arcetri, is not to go out of the house, is not to receive visitors, and that if he asks for more favours, or transgresses the commands laid upon him, he is liable to be haled back to Rome and cast into a dungeon. These harsh measures were dictated, not by cruelty, but by the fear of his still spreading heresy by conversation, and so he was to be kept isolated.

Idle, however, he was not and could not be. He often complains that his head is too busy for his body. In the enforced solitude of Arcetri he was composing those dialogues on motion which are now reckoned his greatest and most solid achievement. In these the true laws of motion are set forth for the first time (see page 167). One more astronomical discovery also he was to make—that of the moon's libration.

And then there came one more crushing blow. His eyes became inflamed and painful—the sight of one of them failed, the other soon went; he became totally blind. But this, being a heaven-sent infliction, he could bear with resignation, though it must have been keenly painful to a solitary man of his activity. "Alas!" says he, in one of his letters, "your dear friend and servant is totally blind. Henceforth this heaven, this universe, which by wonderful observations I had enlarged a hundred and a thousand times beyond the conception of former ages, is shrunk for me into the narrow space which I myself fill in it. So it pleases God; it shall therefore please me also."

He was now allowed an amanuensis, and the help of his pupils Torricelli, Castelli, and Viviani, all devotedly attached to him, and Torricelli very famous after him. Visitors also were permitted, after approval by a Jesuit supervisor; and under these circumstances many visited him, among them a man as immortal as himself—John Milton, then only twenty-nine, travelling in Italy. Surely a pathetic incident, this meeting of these two great men—the one already blind, the other destined to become so. No wonder that, as in his old age he dictated his masterpiece, the thoughts of the English poet should run on the blind sage of Tuscany, and the reminiscence of their conversation should lend colour to the poem.

Well, it were tedious to follow the petty annoyances and troubles to which Galileo was still subject—how his own son was set to see that no unauthorized procedure took place, and that no heretic visitors were admitted; how it was impossible to get his new book printed till long afterwards; and how one form of illness after another took possession of him. The merciful end came at last, and at the age of seventy-eight he was released from the Inquisition.

They wanted to deny him burial—they did deny him a monument; they threatened to cart his bones away from Florence if his friends attempted one. And so they hoped that he and his work might be forgotten.

Poor schemers! Before the year was out an infant was born in Lincolnshire, whose destiny it was to round and complete and carry forward the work of their victim, so that, until man shall cease from the planet, neither the work nor its author shall have need of a monument.

Here might I end, were it not that the same kind of struggle as went on fiercely in the seventeenth century is still smouldering even now. Not in astronomy indeed, as then; nor yet in geology, as some fifty years ago; but in biology mainly—perhaps in other subjects. I myself have heard Charles Darwin spoken of as an atheist and an infidel, the theory of evolution assailed as unscriptural, and the doctrine of the ascent of man from a lower state of being, as opposed to the fall of man from some higher condition, denied as impious and un-Christian.

Men will not learn by the past; still they brandish their feeble weapons against the truths of Nature, as if assertions one way or another could alter fact, or make the thing other than it really is. As Galileo said before his spirit was broken, "In these and other positions certainly no man doubts but His Holiness the Pope hath always an absolute power of admitting or condemning them; but it is not in the power of any creature to make them to be true or false, or otherwise than of their own nature and in fact they are."

I know nothing of the views of any here present; but I have met educated persons who, while they might laugh at the men who refused to look through a telescope lest they should learn something they did not like, yet also themselves commit the very same folly. I have met persons who utterly refuse to listen to any view concerning the origin of man other than that of a perfect primæval pair in a garden, and I am constrained to say this much: Take heed lest some prophet, after having excited your indignation at the follies and bigotry of a bygone generation, does not turn upon you with the sentence, "Thou art the man."

SUMMARY OF FACTS FOR LECTURE VI

Science before Newton

Dr. Gilbert, of Colchester, Physician to Queen Elizabeth, was an excellent experimenter, and made many discoveries in magnetism and electricity. He was contemporary with Tycho Brahé, and lived from 1540 to 1603.

Francis Bacon, Lord Verulam, 1561–1626, though a brilliant writer, is not specially important as regards science. He was not a scientific man, and his rules for making discoveries, or methods of induction, have never been consciously, nor often indeed unconsciously, followed by discoverers. They are not in fact practical rules at all, though they were so intended. His really strong doctrines are that phenomena must be studied direct, and that variations in the ordinary course of nature must be induced by aid of experiment; but he lacked the scientific instinct for pursuing these great truths into detail and special cases. He sneered at the work and methods of both Gilbert and Galileo, and rejected the Copernican theory as absurd. His literary gifts have conferred on him an artificially high scientific reputation, especially in England; at the same time his writings undoubtedly helped to make popular the idea of there being new methods for investigating Nature, and, by insisting on the necessity for freedom from preconceived ideas and opinions, they did much to release men from the bondage of Aristotelian authority and scholastic tradition.

The greatest name between Galileo and Newton is that of Descartes.

René Descartes was born at La Haye in Touraine, 1596, and died at Stockholm in 1650. He did important work in mathematics, physics, anatomy, and philosophy. Was greatest as a philosopher and mathematician. At the age of twenty-one he served as a volunteer under Prince Maurice of Nassau, but spent most of his later life in Holland. His famous *Discourse on Method* appeared at Leyden in 1637, and his *Principia* at Amsterdam in 1644; great pains being taken to avoid the condemnation of the Church.

Descartes's main scientific achievement was the application of algebra to geometry; his most famous speculation was the "theory of vortices," invented to account for the motion of planets. He also made many discoveries in optics and physiology. His best known immediate pupils were the Princess Elizabeth of Bohemia, and Christina, Queen of Sweden.

He founded a distinct school of thought (the Cartesian), and was the precursor of the modern mathematical method of investigating science, just as Galileo and Gilbert were the originators of the modern experimental method.

LECTURE VI

DESCARTES AND HIS THEORY OF VORTICES

AFTER the dramatic life we have been considering in the last two lectures, it is well to have a breathing space, to look round on what has been accomplished, and to review the state of scientific thought, before proceeding to the next great era. For we are still in the early morning of scientific discovery: the dawn of the modern period, faintly heralded by Copernicus, brought nearer by the work of Tycho and Kepler, and introduced by the discoveries of Galileo—the dawn has occurred, but the sun is not yet visible. It is hidden by the clouds and mists of the long night of ignorance and prejudice. The light is sufficient, indeed, to render these earth-born vapours more visible: it is not sufficient to dispel them. A generation of slow and doubtful progress must pass, before the first ray of sunlight can break through the eastern clouds and the full orb of day itself appear.

It is this period of hesitating progress and slow leavening of men's ideas that we have to pass through in this week's lecture. It always happens thus: the assimilation of great and new ideas is always a slow and gradual process: there is no haste either here or in any other department of Nature. *Die Zeit ist unendlich lang.* Steadily the forces work, sometimes seeming to accomplish nothing; sometimes even the motion appears retrograde; but in the long run the destined end is reached, and the course, whether of a planet or of men's thoughts about the universe, is permanently altered. Then, the controversy was about the *earth's* place in the universe; now, if there be any controversy of the same kind, it is about *man's* place in the universe; but the process is the same: a startling statement by a great genius or prophet, general disbelief, and, it may be, an attitude of hostility, gradual acceptance by a few, slow spreading among the many, ending in universal acceptance and faith often as unquestioning and unreasoning as the old state of unfaith had been. Now the process is comparatively speedy: twenty years accomplishes a great deal: then it was tediously slow, and a century seemed to accomplish very little. Periodical literature may be responsible for some waste of time, but it certainly assists the rapid spread of ideas. The rate with which ideas are assimilated by the general public cannot even now be considered excessive, but how much faster it is than it was a few centuries ago may be illustrated by the attitude of the public to Darwinism now, twenty-five years after *The Origin of Species*, as compared with their attitude to the Copernican system a century after *De Revolutionibus*. By the way, it is, I know, presumptuous for me to have an opinion, but I cannot hear Darwin compared to or mentioned along with Newton without a shudder. The stage in which he found biology seems to me far more comparable with the Ptolemaic era in astronomy, and he himself to be quite fairly comparable to Copernicus.

Let us proceed to summarize the stage at which the human race had arrived at the epoch with which we are now dealing.

The Copernican view of the solar system had been stated, restated, fought, and insisted on; a chain of brilliant telescopic discoveries had made it popular and accessible to all men of any intelligence: henceforth it must be left to slowly percolate and sink into the minds of the people. For the nations were waking up now, and were accessible to new ideas. England especially was, in some sort, at the zenith of its glory; or, if not at the zenith, was in that full flush of youth and expectation and hope which is stronger and more prolific of great deeds and thoughts than a maturer period.

A common cause against a common and detested enemy had roused in the hearts of Englishmen a passion of enthusiasm and patriotism; so that the mean elements of trade, their cheating yard-wands, were forgotten for a time; the Armada was defeated, and the nation's true and conscious adult life began. Commerce was now no mere struggle for profit and hard bargains; it was full of the spirit of adventure and discovery; a new world had been opened up; who could tell what more remained unexplored? Men awoke to the splendour of their inheritance, and away sailed Drake and Frobisher and Raleigh into the lands of the West.

For literature, you know what a time it was. The author of *Hamlet* and *Othello* was alive: it is needless to say more. And what about science? The atmosphere of science is a more quiet and less stirring one; it thrives best when the fever of excitement is allayed; it is necessarily a later growth than literature. Already, however, our second great man of science was at work in a quiet country town—second in point of time, I mean, Roger Bacon being the first. Dr. Gilbert, of Colchester, was the second in point of time, and the age was ripening for the time when England was to be honoured with such a galaxy of scientific luminaries—Hooke and Boyle and Newton—as the world had not yet known.

Yes, the nations were awake. "In all directions," as Draper says, "Nature was investigated: in all directions new methods of examination were yielding unexpected and beautiful results. On the ruins of its ivy-grown cathedrals Ecclesiasticism [or Scholasticism], surprised and blinded by the breaking day, sat solemnly blinking at the light and life about it, absorbed in the recollection of the night that had passed, dreaming of new phantoms and delusions in its wished-for return, and vindictively striking its talons at any derisive assailant who incautiously approached too near."

Of the work of Gilbert there is much to say; so there is also of Roger Bacon, whose life I am by no means sure I did right in omitting. But neither of them had much to do with astronomy, and since it is in astronomy that the most startling progress was during these centuries being made, I have judged it wiser to adhere mainly to the pioneers in this particular department.

Only for this reason do I pass Gilbert with but slight mention. He knew of the Copernican theory and thoroughly accepted it (it is convenient to speak of it as the Copernican theory, though you know that it had been considerably improved in detail since the first crude statement by Copernicus), but he made in it no changes. He was a cultivated scientific man, and an acute experimental philosopher; his main work lay in the domain of magnetism and electricity. The phenomena connected with the mariner's compass had been studied somewhat by Roger Bacon; and they were now examined still more thoroughly by Gilbert, whose treatise *De Magnete*, marks the beginning of the science of magnetism.

As an appendix to that work he studied the phenomenon of amber, which had been mentioned by Thales. He resuscitated this little fact after its burial of 2,200 years, and greatly extended it. He it was who invented the name electricity—I wish it had been a shorter one. Mankind invents names much better than do philosophers. What can be better than "heat," "light," "sound"? How favourably they compare with electricity, magnetism, galvanism, electro-magnetism, and magneto-electricity! The only long-established monosyllabic name I know invented by a philosopher is "gas"—an excellent attempt, which ought to be imitated.[12]

Of Lord Bacon, who flourished about the same time (a little later), it is necessary to say something, because many persons are under the impression that to him and his *Novum Organon* the reawakening of the world, and the overthrow of Aristotelian tradition, are mainly due. His influence, however, has been exaggerated. I am not going to enter into a discussion of the *Novum Organon*, and the mechanical methods which he propounded as certain to evolve truth if patiently pursued; for this is what he thought he was doing—giving to the world an infallible recipe for discovering truth, with which any ordinarily industrious man could make discoveries by means of collection and discrimination of instances. You will take my statement for what it is worth, but I assert this: that many of the methods which Bacon lays down are not those which the experience of mankind has found to be serviceable; nor are they such as a scientific man would have thought of devising.

True it is that a real love and faculty for science are born in a man, and that to the man of scientific capacity rules of procedure are unnecessary; his own intuition is sufficient, or he has mistaken his vocation,—but that is not my point. It is not that Bacon's methods are useless because the best men do not need them; if they had been founded on a careful study of the methods actually employed, though it might be unconsciously employed, by scientific men—as the methods of induction, stated long after by John Stuart Mill, were founded—then, no doubt, their statement would have been a valuable service and a great thing to accomplish. But they were not this. They are the ideas of a brilliant man of letters, writing in an age when scientific research was almost unknown, about a subject in which he was an amateur. I confess I do not see how he, or John Stuart Mill, or any one else, writing in that age, could have formulated the true rules of philosophizing; because the materials and information were scarcely to hand. Science and its methods were only beginning to grow. No doubt it was a brilliant attempt. No doubt also there are many good and true points in the statement, especially in his insistence on the attitude of free and open candour with which the investigation of Nature should be approached. No doubt there was much beauty in his allegories of the errors into which men were apt to fall—the *idola* of the market-place, of the tribe, of the theatre, and of the den; but all this is literature, and on the solid progress of science may be said to have had little or no effect. Descartes's *Discourse on Method* was a much more solid production.

You will understand that I speak of Bacon purely as a scientific man. As a man of letters, as a lawyer, a man of the world, and a statesman, he is beyond any criticism of mine. I speak only of the purely scientific aspect of the *Novum Organon*. *The Essays* and *The Advancement of Learning* are masterly productions; and as a literary man he takes high rank.

The over-praise which, in the British Isles, has been lavished upon his scientific importance is being followed abroad by what may be an unnecessary amount of detraction. This is always the worst of setting up a man on too high a pinnacle; some one has to undertake the ungrateful task of pulling him down again. Justus von Liebig addressed himself to this task with some vigour in his *Reden und Abhandlung* (Leipzig, 1874), where he quotes from Bacon a number of suggestions for absurd experimentation.[13]

The next paragraph I read, not because I endorse it, but because it is always well to hear both sides of a question. You have probably been long accustomed to read over-estimates of Bacon's importance, and extravagant laudation of his writings as making an epoch in science; hear what Draper says on the opposite side:—[14]

"The more closely we examine the writings of Lord Bacon, the more unworthy does he seem to have been of the great reputation which has been awarded to him. The popular delusion to which he owes so much originated at a time when the history of science was unknown. They who first brought him into notice knew nothing of the old school of Alexandria. This boasted founder of a new philosophy could not comprehend, and would not accept, the greatest of all scientific doctrines when it was plainly set before his eyes.

"It has been represented that the invention of the true method of physical science was an amusement of Bacon's hours of relaxation from the more laborious studies of law, and duties of a Court.

"His chief admirers have been persons of a literary turn, who have an idea that scientific discoveries are accomplished by a mechanico-mental operation. Bacon never produced any great practical result himself, no great physicist has ever made any use of his method. He has had the same to do with the development of modern science that the inventor of the orrery has had to do with the discovery of the mechanism of the world. Of all the important physical discoveries, there is not one which shows that its author made it by the Baconian instrument.

"Newton never seems to have been aware that he was under any obligation to Bacon. Archimedes, and the Alexandrians, and the Arabians, and Leonardo da Vinci did very well before he was born; the discovery of America by Columbus and the circumnavigation by Magellan can hardly be attributed to him, yet they were the consequences of a truly philosophical reasoning. But the investigation of Nature is an affair of genius, not of rules. No man can invent an *organon* for writing tragedies and epic poems. Bacon's system is, in its own terms, an idol of the theatre. It would scarcely guide a man to a solution of the riddle of Ælia Lælia Crispis, or to that of the charade of Sir Hilary.

"Few scientific pretenders have made more mistakes than Lord Bacon. He rejected the Copernican system, and spoke insolently of its great author; he undertook to criticize adversely Gilbert's treatise *De Magnete*; he was occupied in the condemnation of any investigation of final causes, while Harvey was deducing the circulation of the blood from Aquapendente's discovery of the valves in the veins; he was doubtful whether instruments were of any advantage, while Galileo was investigating the heavens with the telescope. Ignorant himself of every branch of mathematics, he presumed that they were useless in science but a few years before Newton achieved by their aid his immortal discoveries.

"It is time that the sacred name of philosophy should be severed from its long connection with that of one who was a pretender in science, a time-serving politician, an insidious lawyer, a corrupt judge, a treacherous friend, a bad man."

This seems to me a depreciation as excessive as are the eulogies commonly current. The truth probably lies somewhere between the two extremes. It is unfair to judge Bacon's methods by thinking of physical science in its present stage. To realise his position we must think of a subject still in its very early infancy, one in which the advisability of applying experimental methods is still doubted; one which has been studied by means of books and words and discussion of normal instances, instead of by collection and observation of the unusual and irregular, and by experimental production of variety. If we think of a subject still in this infantile and almost pre-scientific stage, Bacon's words and formulæ are far from inapplicable; they are, within their limitations, quite necessary and wholesome. A subject in this stage, strange to say, exists,—psychology; now hesitatingly beginning to assume its experimental weapons amid a stifling atmosphere of distrust and suspicion. Bacon's lack of the modern scientific instinct must be admitted, but he rendered humanity a powerful service in directing it from books to nature herself, and his genius is indubitable. A judicious account of his life and work is given by Prof. Adamson, in the *Encyclopædia Britannica*, and to this article I now refer you.

Who, then, was the man of first magnitude filling up the gap in scientific history between the death of Galileo and the maturity of Newton? Unknown and mysterious are the laws regulating the appearance of genius. We have passed in review a Pole, a Dane, a German, and an Italian,—the great man is now a Frenchman, René Descartes, born in Touraine, on the 31st of March, 1596.

His mother died at his birth; the father was of no importance, save as the owner of some landed property. The boy was reared luxuriously, and inherited a fair fortune. Nearly all the men of first rank, you notice, were born well off. Genius born to poverty might, indeed, even then achieve name and fame—as we see in the case of Kepler—but it was terribly handicapped. Handicapped it is still, but far less than of old; and we may hope it will become gradually still less so as enlightenment proceeds, and the tremendous moment of great men to a nation is more clearly and actively perceived.

It is possible for genius, when combined with strong character, to overcome all obstacles, and reach the highest eminence, but the struggle must be severe; and the absence of early training and refinement during the receptive years of youth must be a lifelong drawback.

Descartes had none of these drawbacks; life came easily to him, and, as a consequence perhaps, he never seems to have taken it quite seriously. Great movements and stirring events were to him opportunities for the study of men and manners; he was not the man to court persecution, nor to show enthusiasm for a losing or struggling cause.

In this, as in many other things, he was imbued with a very modern spirit, a cynical and sceptical spirit, which, to an outside and superficial observer like myself, seems rather rife just now.

He was also imbued with a phase of scientific spirit which you sometimes still meet with, though I believe it is passing away, viz. an uncultured absorption in his own pursuits, and some feeling of contempt for classical and literary and æsthetic studies.

In politics, art, and history he seems to have had no interest. He was a spectator rather than an actor on the stage of the world; and though he joined the army of that great military commander Prince Maurice of Nassau, he did it not as a man with a cause at heart worth fighting for, but precisely in the spirit in which one of our own gilded youths would volunteer in a similar case, as a good opportunity for frolic and for seeing life.

He soon tired of it and withdrew—at first to gay society in Paris. Here he might naturally have sunk into the gutter with his companions, but for a great mental shock which became the main epoch and turning-point of his life, the crisis which diverted him from frivolity to seriousness. It was a purely intellectual emotion, not excited by anything in the visible or tangible world; nor could it be called conversion in the common acceptation of that term. He tells us that on the 10th of November, 1619, at the age of twenty-four, a brilliant idea flashed upon him—the first idea, namely, of his great and powerful mathematical method, of which I will speak directly; and in the flush of it he foresaw that just as geometers, starting with a few simple and evident propositions or axioms, ascend by a long and intricate ladder of reasoning to propositions more and more abstruse, so it might be possible to ascend from a few data, to all the secrets and facts of the universe, by a process of mathematical reasoning.

"Comparing the mysteries of Nature with the laws of mathematics, he dared to hope that the secrets of both could be unlocked with the same key."

That night he lapsed gradually into a state of enthusiasm, in which he saw three dreams or visions, which he interpreted at the time, even before waking, to be revelations from the Spirit of Truth to direct his future course, as well as to warn him from the sins he had already committed.

His account of the dreams is on record, but is not very easy to follow; nor is it likely that a man should be able to convey to others any adequate idea of the deepest spiritual or mental agitation which has shaken him to his foundations.

His associates in Paris were now abandoned, and he withdrew, after some wanderings, to Holland, where he abode the best part of his life and did his real work.

Even now, however, he took life easily. He recommends idleness as necessary to the production of good mental work. He worked and meditated but a few hours a day: and most of those in bed. He used to think best in bed, he said. The afternoon he devoted to society and recreation. After supper he wrote letters to various persons, all plainly intended for publication, and scrupulously preserved. He kept himself free from care, and was most cautious about his health, regarding himself, no doubt, as a subject of experiment, and wishful to see how long he could prolong his life. At one time he writes to a friend that he shall be seriously disappointed if he does not manage to see 100 years.

FIG. 53.—Descartes.

This plan of not over-working himself, and limiting the hours devoted to serious thought, is one that might perhaps advantageously be followed by some over-laborious students of the present day. At any rate it conveys a lesson; for the amount of ground covered by Descartes, in a life not very long, is extraordinary. He must, however, have had a singular aptitude for scientific work; and the judicious leaven of selfishness whereby he was able to keep himself free from care and embarrassments must have been a great help to him.

And what did his versatile genius accomplish during his fifty-four years of life?

In philosophy, using the term as meaning mental or moral philosophy and metaphysics, as opposed to natural philosophy or physics, he takes a very high rank, and it is on this that perhaps his greatest fame rests. (He is the author, you may remember, of the famous aphorism, "*Cogito, ergo sum.*")

In biology I believe he may be considered almost equally great: certainly he spent a great deal of time in dissecting, and he made out a good deal of what is now known of the structure of the body, and of the theory of vision. He eagerly accepted the doctrine of the circulation of the blood, then being taught by Harvey, and was an excellent anatomist.

You doubtless know Professor Huxley's article on Descartes in the *Lay Sermons*, and you perceive in what high estimation he is there held.

He originated the hypothesis that animals are automata, for which indeed there is much to be said from some points of view; but he unfortunately believed that they were unconscious and non-sentient automata, and this belief led his disciples into acts of abominable cruelty. Professor Huxley lectured on this hypothesis and partially upheld it not many years since. The article is included in his volume called *Science and Culture.*

Concerning his work in mathematics and physics I can speak with more confidence. He is the author of the Cartesian system of algebraic or analytic geometry, which has been so powerful an engine of research, far easier to wield than the old synthetic geometry. Without it Newton could never have written the *Principia*, or made his greatest discoveries. He might indeed have invented it for himself, but it would have consumed some of his life to have brought it to the necessary perfection.

The principle of it is the specification of the position of a point in a plane by two numbers, indicating say its distance from two lines of reference in the plane; like the latitude and longitude of a place on the globe. For instance, the two lines of reference might be the bottom edge and the left-hand vertical edge of a wall; then a point on the wall, stated as being for instance 6 feet along and 2 feet up, is precisely determined. These two distances are called co-ordinates; horizontal ones are usually denoted by x, and vertical ones by y.

If, instead of specifying two things, only one statement is made, such as $y = 2$, it is satisfied by a whole row of points, all the points in a horizontal line 2 feet above the ground. Hence $y = 2$ may be said to represent that straight line, and is called the equation to that straight line. Similarly $x = 6$ represents a vertical straight line 6 feet (or inches or some other unit) from the left-hand edge. If it is asserted that $x = 6$ and $y = 2$, only one point can be found to satisfy both conditions, viz. the crossing point of the above two straight lines.

Suppose an equation such as $x = y$ to be given. This also is satisfied by a row of points, viz. by all those that are equidistant from bottom and left-hand edges. In other words, $x = y$ represents a straight line slanting upwards at 45°. The equation $x = 2y$ represents another straight line with a different angle of slope, and so on. The equation $x^2 + y^2 = 36$ represents a circle of radius 6. The equation $3x^2 + 4y^2 = 25$ represents an ellipse; and in general every algebraic equation that can be written down, provided it involve only two variables, x and y, represents some curve in a plane; a curve moreover that can be drawn, or its properties completely investigated without drawing, from the equation. Thus algebra is wedded to geometry, and the investigation of geometric relations by means of algebraic equations is called analytical geometry, as opposed to the old Euclidian or synthetic mode of treating the subject by reasoning consciously directed to the subject by help of figures.

If there be three variables—x, y, and z,—instead of only two, an equation among them represents not a curve in a plane but a surface in space; the three variables corresponding to the three dimensions of space: length, breadth, and thickness.

An equation with four variables usually requires space of four dimensions for its geometrical interpretation, and so on.

Thus geometry can not only be reasoned about in a more mechanical and therefore much easier, manner, but it can be extended into regions of which we have and can have no direct conception, because we are deficient in sense organs for accumulating any kind of experience in connexion with such ideas.

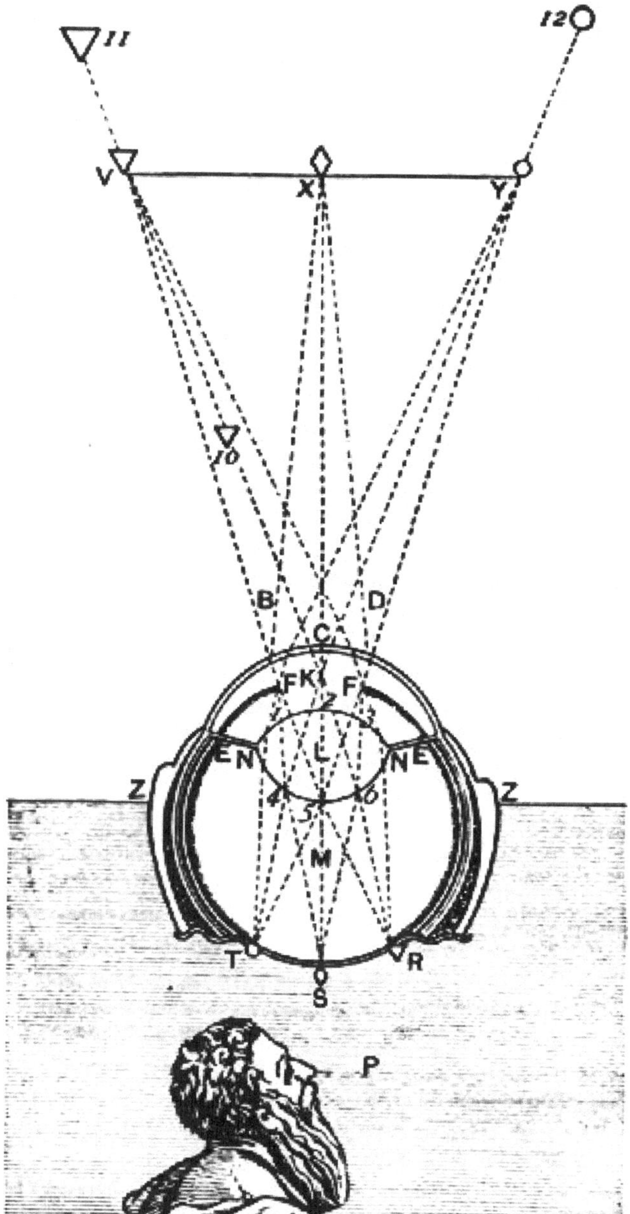

FIG. 54.—The eye diagram. [From Descartes' *Principia.*] Three external points are shown depicted on the retina: the image being appreciated by a representation of the brain.

In physics proper Descartes' tract on optics is of considerable historical interest. He treats all the subjects he takes up in an able and original manner.

In Astronomy he is the author of that famous and long upheld theory, the doctrine of vortices.

He regarded space as a plenum full of an all-pervading fluid. Certain portions of this fluid were in a state of whirling motion, as in a whirlpool or eddy of water; and each planet had its own eddy, in which it was whirled round and round, as a straw is caught and whirled in a common whirlpool. This idea he works out and elaborates very fully, applying it to the system of the world, and to the explanation of all the motions of the planets.

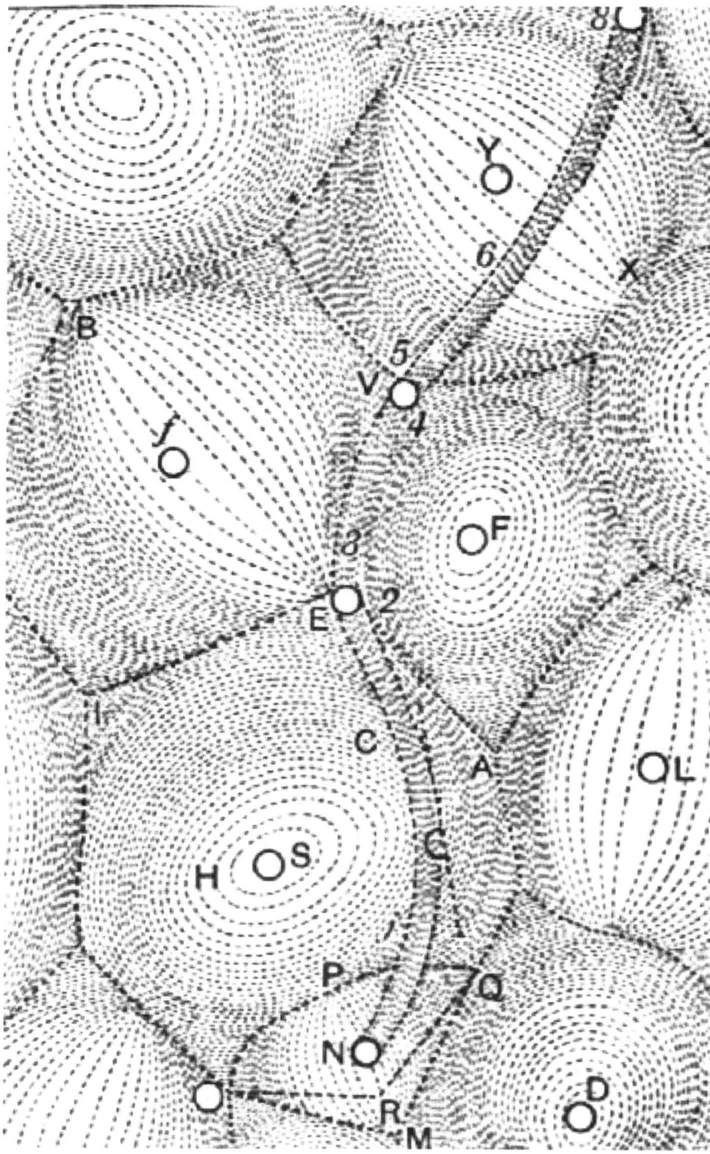

FIG. 55.—Descartes's diagram of vortices, from his *Principia*.

This system evidently supplied a void in men's minds, left vacant by the overthrow of the Ptolemaic system, and it was rapidly accepted. In the English Universities it held for a long time almost undisputed sway; it was in this faith that Newton was brought up.

Something was felt to be necessary to keep the planets moving on their endless round; the *primum mobile* of Ptolemy had been stopped; an angel was sometimes assigned to each planet to carry it round, but though a widely diffused belief, this was a fantastic and not a serious scientific one. Descartes's vortices seemed to do exactly what was wanted.

It is true they had no connexion with the laws of Kepler. I doubt whether he knew about the laws of Kepler; he had not much opinion of other people's work; he read very little—found it easier to think. (He travelled through Florence once when Galileo was at the height of his renown without calling upon or seeing him.) In so far as the motion of a planet was not circular, it had to be accounted for by the jostling and crowding and distortion of the vortices.

Gravitation he explained by a settling down of bodies toward the centre of each vortex; and cohesion by an absence of relative motion tending to separate particles of matter. He "can imagine no stronger cement."

The vortices, as Descartes imagined them, are not now believed in. Are we then to regard the system as absurd and wholly false? I do not see how we can do this, when to this day philosophers are agreed in believing space to be completely full of fluid, which fluid is certainly capable of vortex motion, and perhaps everywhere does possess that motion. True, the now imagined vortices are not the large whirls of planetary size, they are rather infinitesimal whirls of less than atomic dimensions; still a whirling fluid is believed in to this day, and many are seeking to deduce all the properties of matter (rigidity, elasticity, cohesion gravitation, and the rest) from it.

Further, although we talk glibly about gravitation and magnetism, and so on, we do not really know what they are. Progress is being made, but we do not yet properly know. Much, overwhelmingly much, remains to be discovered, and it ill-behoves us to reject any well-founded and long-held theory as utterly and intrinsically false and absurd. The more one gets to know, the more one perceives a kernel of truth even in the most singular statements; and scientific men have learned by experience to be very careful how they lop off any branch of the tree of knowledge, lest as they cut away the dead wood they lose also some green shoot, some healthy bud of unperceived truth.

However, it may be admitted that the idea of a Cartesian vortex in connexion with the solar system applies, if at all, rather to an earlier—its nebulous—stage, when the whole thing was one great whirl, ready to split or shrink off planetary rings at their appropriate distances.

Soon after he had written his great work, the *Principia Mathematica*, and before he printed it, news reached him of the persecution and recantation of Galileo. "He seems to have been quite thunderstruck at the tidings," says Mr. Mahaffy, in his *Life of Descartes*.[15] "He had started on his scientific journeys with the firm determination to enter into no conflict with the Church, and to carry out his system of pure mathematics and physics without ever meddling with matters of faith. He was rudely disillusioned as to the possibility of this severance. He wrote at once—apparently, November 20th, 1633—to Mersenne to say he would on no account publish his work—nay, that he had at first resolved to burn all his papers, for that he would never prosecute philosophy at the risk of being censured by his Church. 'I could hardly have believed,' he says, 'that an Italian, and in favour with the Pope as I hear, could be considered criminal for nothing else than for seeking to establish the earth's motion; though I know it has formerly been censured by some Cardinals. But I thought I had heard that since then it was constantly being taught, even at Rome; and I confess that if the opinion of the earth's movement is false, all the foundations of my philosophy are so also, because it is demonstrated clearly by them. It is so bound up with every part of my treatise that I could not sever it without making the remainder faulty; and although I consider all my conclusions based on very certain and clear demonstrations, I would not for all the world sustain them against the authority of the Church.'"

Ten years later, however, he did publish the book, for he had by this time hit on an ingenious compromise. He formally denied that the earth moved, and only asserted that it was carried along with its water and air in one of those larger motions of the celestial ether which produce the diurnal and annual revolutions of the solar system. So, just as a passenger on the deck of a ship might be called stationary, so was the earth. He gives himself out therefore as a follower of Tycho rather than of Copernicus, and says if the Church won't accept this compromise he must return to the Ptolemaic system; but he hopes they won't compel him to do that, seeing that it is manifestly untrue.

This elaborate deference to the powers that be did not indeed save the work from being ultimately placed upon the forbidden list by the Church, but it saved himself, at any rate, from annoying persecution. He was not, indeed, at all willing to be persecuted, and would no doubt have at once withdrawn anything they wished. I should be sorry to call him a time-server, but he certainly had plenty of that worldly wisdom in which some of his predecessors had been so lamentably deficient. Moreover, he was really a sceptic, and cared nothing at all about the Church or its dogmas. He knew the Church's power, however, and the advisability of standing well with it: he therefore professed himself a Catholic, and studiously kept his science and his Christianity distinct.

In saying that he was a sceptic you must not understand that he was in the least an atheist. Very few men are; certainly Descartes never thought of being one. The term is indeed ludicrously inapplicable to him, for a great part of his philosophy is occupied with what he considers a rigorous proof of the existence of the Deity.

At the age of fifty-three he was sent for to Stockholm by Christina, Queen of Sweden, a young lady enthusiastically devoted to study of all kinds and determined to surround her Court with all that was most famous in literature and science. Thither, after hesitation, Descartes went. He greatly liked royalty, but he dreaded the cold climate. Born in Touraine, a Swedish winter was peculiarly trying to him, especially as the energetic Queen would have lessons given her at five o'clock in the morning. She intended to treat him well, and was immensely taken with him; but this getting up at five o'clock on a November morning, to a man accustomed all his life to lie in bed till eleven, was a cruel hardship. He was too much of a courtier, however, to murmur, and the early morning audience continued. His health began to break down: he thought of retreating, but suddenly he gave way and became delirious. The Queen's physician attended him, and of course wanted to bleed him. This, knowing all he knew of physiology, sent him furious, and they could do nothing with him. After some days he became quiet, was bled twice, and gradually sank, discoursing with great calmness on his approaching death, and duly fortified with all the rites of the Catholic Church.

His general method of research was as nearly as possible a purely deductive one:—*i.e.*, after the manner of Euclid he starts with a few simple principles, and then, by a chain of reasoning, endeavours to deduce from them their consequences, and so to build up bit by bit an edifice of connected knowledge. In this he was the precursor of Newton. This method, when rigorously pursued, is the most powerful and satisfactory of all, and results in an ordered province of science far superior to the fragmentary conquests of experiment. But few indeed are the men who can handle it safely and satisfactorily: and none without continual appeals to experiment for verification. It was through not perceiving the necessity for verification that he erred. His importance to science lies not so much in what he actually discovered as in his anticipation of the right conditions for the solution of problems in physical science. He in fact made the discovery that Nature could after all be interrogated mathematically—a fact that was in great danger of remaining unknown. For, observe, that the mathematical study of Nature, the discovery of truth with a piece of paper and a pen, has a perilous similarity at first sight to the straw-thrashing subtleties of the Greeks, whose methods of investigating nature by discussing the meaning of words and the usage of language and the necessities of thought, had proved to be so futile and unproductive.

A reaction had set in, led by Galileo, Gilbert, and the whole modern school of experimental philosophers, lasting down to the present day:—men who teach that the only right way of investigating Nature is by experiment and observation.

It is indeed a very right and an absolutely necessary way; but it is not the only way. A foundation of experimental fact there must be; but upon this a great structure of theoretical deduction can be based, all rigidly connected together by pure reasoning, and all necessarily as true as the premises, provided no mistake is made. To guard against the possibility of mistake and oversight, especially oversight, all conclusions must sooner or later be brought to the test of experiment; and if disagreeing therewith, the theory itself must be re-examined, and the flaw discovered, or else the theory must be abandoned.

Of this grand method, quite different from the gropings in the dark of Kepler—this method, which, in combination with experiment, has made science what it now is—this which in the hands of Newton was to lead to such stupendous results, we owe the beginning and early stages to René Descartes.

SUMMARY OF FACTS FOR LECTURES VII AND VIII

Otto Guericke	1602–1686
Hon. Robert Boyle	1626–1691
Huyghens	1629–1695
Christopher Wren	1632–1723
Robert Hooke	1635–1702
NEWTON	1642–1727
Edmund Halley	1656–1742
James Bradley	1692–1762

Chronology of Newton's Life.

Isaac Newton was born at Woolsthorpe, near Grantham, Lincolnshire, on Christmas Day, 1642. His father, a small freehold farmer, also named Isaac, died before his birth. His mother, *née* Hannah Ayscough, in two years married a Mr. Smith, rector of North Witham, but was again left a widow in 1656. His uncle, W. Ayscough, was rector of a near parish and a graduate of Trinity College, Cambridge. At the age of fifteen Isaac was removed from school at Grantham to be made a farmer of, but as it seemed he would not make a good one his uncle arranged for him to return to school and thence to Cambridge, where he entered Trinity College as a sub-sizar in 1661. Studied Descartes's geometry. Found out a method of infinite series in 1665, and began the invention of Fluxions. In the same year and the next he was driven from Cambridge by the plague. In 1666, at Woolsthorpe, the apple fell. In 1667 he was elected a fellow of his college, and in 1669 was specially noted as possessing an unparalleled genius by Dr. Barrow, first Lucasian Professor of Mathematics. The same year Dr. Barrow retired from his chair in favour of Newton, who was thus elected at the age of twenty-six. He lectured first on optics with great success. Early in 1672 he was elected a Fellow of the Royal Society, and communicated his researches in optics, his reflecting telescope, and his discovery of the compound nature of white light. Annoying controversies arose; but he nevertheless contributed a good many other most important papers in optics, including observations in diffraction, and colours of thin plates. He also invented the modern sextant. In 1672 a letter from Paris was read at the Royal Society concerning a new and accurate determination of the size of the earth by Picard. When Newton heard of it he began the *Principia*, working in silence. In 1684 arose a discussion between Wren, Hooke, and Halley concerning the law of inverse square as applied to gravity and the path it would cause the planets to describe. Hooke asserted that he had a solution, but he would not produce it. After waiting some time for it Halley went to Cambridge to consult Newton on the subject, and thus discovered the existence of the first part of the *Principia*, wherein all this and much more was thoroughly worked out. On his representations to the Royal Society the manuscript was asked for, and when complete was printed and published in 1687 at Halley's expense. While it was being completed Newton and seven others were sent to uphold the dignity of the University, before the Court of High Commission and Judge Jeffreys, against a high-handed action of James II. In 1682 he was sent to Parliament, and was present at the coronation of William and Mary. Made friends with Locke. In 1694 Montague, Lord Halifax, made him Warden, and in 1697 Master, of the Mint. Whiston succeeded him as Lucasian Professor. In 1693 the method of fluxions was published. In 1703 Newton was made President of the Royal Society, and held the office to the end of his life. In 1705 he was knighted by Anne. In 1713 Cotes helped him to bring out a new edition of the *Principia*, completed as we now have it. On the 20th of March 1727, he died: having lived from Charles I. to George II.

THE LAWS OF MOTION, DISCOVERED BY GALILEO, STATED BY NEWTON.

Law 1.—If no force acts on a body in motion, it continues to move uniformly in a straight line.

Law 2.—If force acts on a body, it produces a change of motion proportional to the force and in the same direction.

Law 3.—When one body exerts force on another, that other reacts with equal force upon the one.

LECTURE VII

SIR ISAAC NEWTON

THE little hamlet of Woolsthorpe lies close to the village of Colsterworth, about six miles south of Grantham, in the county of Lincoln. In the manor house of Woolsthorpe, on Christmas Day, 1642, was born to a widowed mother a sickly infant who seemed not long for this world. Two women who were sent to North Witham to get some medicine for him scarcely expected to find him alive on their return. However, the child lived, became fairly robust, and was named Isaac, after his father. What sort of a man this father was we do not know. He was what we may call a yeoman, that most wholesome and natural of all classes. He owned the soil he tilled, and his little estate had already been in the family for some hundred years. He was thirty-six when he died, and had only been married a few months.

Of the mother, unfortunately, we know almost as little. We hear that she was recommended by a parishioner to the Rev. Barnabas Smith, an old bachelor in search of a wife, as "the widow Newton—an extraordinary good woman:" and so I expect she was, a thoroughly sensible, practical, homely, industrious, middle-class, Mill-on-the-Floss sort of woman. However, on her second marriage she went to live at North Witham, and her mother, old Mrs. Ayscough, came to superintend the farm at Woolsthorpe, and take care of young Isaac.

By her second marriage his mother acquired another piece of land, which she settled on her first son; so Isaac found himself heir to two little properties, bringing in a rental of about £80 a year.

FIG. 56.—Manor-house of Woolsthorpe.

He had been sent to a couple of village schools to acquire the ordinary accomplishments taught at those places, and for three years to the grammar school at Grantham, then conducted by an old gentleman named Mr. Stokes. He had not been very industrious at school, nor did he feel keenly the fascinations of the Latin Grammar, for he tells us that he was the last boy in the lowest class but one. He used to pay much more attention to the construction of kites and windmills and waterwheels, all of which he made to work very well. He also used to tie paper lanterns to the tail of his kite, so as to make the country folk fancy they saw a comet, and in general to disport himself as a boy should.

It so happened, however, that he succeeded in thrashing, in fair fight, a bigger boy who was higher in the school, and who had given him a kick. His success awakened a spirit of emulation in other things than boxing, and young Newton speedily rose to be top of the school.

Under these circumstances, at the age of fifteen, his mother, who had now returned to Woolsthorpe, which had been rebuilt, thought it was time to train him for the management of his land, and to make a farmer and grazier of him. The boy was doubtless glad to get away from school, but he did not take kindly to the farm—especially not to the marketing at Grantham. He and an old servant were sent to Grantham every week to buy and sell produce, but young Isaac used to leave his old mentor to do all the business, and himself retire to an attic in the house he had lodged in when at school, and there bury himself in books.

After a time he didn't even go through the farce of visiting Grantham at all; but stopped on the road and sat under a hedge, reading or making some model, until his companion returned.

We hear of him now in the great storm of 1658, the storm on the day Cromwell died, measuring the force of the wind by seeing how far he could jump with it and against it. He also made a water-clock and set it up in the house at Grantham, where it kept fairly good time so long as he was in the neighbourhood to look after it occasionally.

At his own home he made a couple of sundials on the side of the wall (he began by marking the position of the sun by the shadow of a peg driven into the wall, but this gradually developed into a regular dial) one of which remained of use for some time; and was still to be seen in the same place during the first half of the present century, only with the gnomon gone. In 1844 the stone on which it was carved was carefully extracted and presented to the Royal Society, who preserve it in their library. The letters WTON roughly carved on it are barely visible.

All these pursuits must have been rather trying to his poor mother, and she probably complained to her brother, the rector of Burton Coggles: at any rate this gentleman found master Newton one morning under a hedge when he ought to have been farming. But as he found him working away at mathematics, like a wise man he persuaded his sister to send the boy back to school for a short time, and then to Cambridge. On the day of his finally leaving school old Mr. Stokes assembled the boys, made them a speech in praise of Newton's character and ability, and then dismissed him to Cambridge.

At Trinity College a new world opened out before the country-bred lad. He knew his classics passably, but of mathematics and science he was ignorant, except through the smatterings he had picked up for himself. He devoured a book on logic, and another on Kepler's Optics, so fast that his attendance at lectures on these subjects became unnecessary. He also got hold of a Euclid and of Descartes's Geometry. The Euclid seemed childishly easy, and was thrown aside, but the Descartes baffled him for a time. However, he set to it again and again and before long mastered it. He threw himself heart and soul into mathematics, and very soon made some remarkable discoveries. First he discovered the binomial theorem: familiar now to all who have done any algebra, unintelligible to others, and therefore I say nothing about it. By the age of twenty-one or two he had begun his great mathematical discovery of infinite series and fluxions—now known by the name of the Differential Calculus. He wrote these things out and must have been quite absorbed in them, but it never seems to have occurred to him to publish them or tell any one about them.

In 1664 he noticed some halos round the moon, and, as his manner was, he measured their angles—the small ones 3 and 5 degrees each, the larger one 22°·35. Later he gave their theory.

Small coloured halos round the moon are often seen, and are said to be a sign of rain. They are produced by the action of minute globules of water or cloud particles upon light, and are brightest when the particles are nearly equal in size. They are not like the rainbow, every part of which is due to light that has entered a raindrop, and been refracted and reflected with prismatic separation of colours; a halo is caused by particles so small as to be almost comparable with the size of waves of light, in a way which is explained in optics under the head "diffraction." It may be easily imitated by dusting an ordinary piece of window-glass over with lycopodium, placing a candle near it, and then looking at the candle-flame through the dusty glass from a fair distance. Or you may look at the image of a candle in a dusted looking-glass. Lycopodium dust is specially suitable, for its granules are remarkably equal in size. The large halo, more rarely seen, of angular radius 22°·35, is due to another cause again, and is a prismatic effect, although it exhibits hardly any colour. The angle 22½° is characteristic of refraction in crystals with angles of 60° and refractive index about the same as water; in other words this halo is caused by ice crystals in the higher regions of the atmosphere.

He also the same year observed a comet, and sat up so late watching it that he made himself ill. By the end of the year he was elected to a scholarship and took his B.A. degree. The order of merit for that year never existed or has not been kept. It would have been interesting, not as a testimony to Newton, but to the sense or non-sense of the examiners. The oldest Professorship of Mathematics at the University of Cambridge, the Lucasian, had not then been long founded, and its first occupant was Dr. Isaac Barrow, an eminent mathematician, and a kind old man. With him Newton made good friends, and was helpful in preparing a treatise on optics for the press. His help is acknowledged by Dr. Barrow in the preface, which states that he had corrected several errors and made some capital additions of his own. Thus we see that, although the chief part of his time was devoted to mathematics, his attention was already directed to both optics and astronomy. (Kepler, Descartes, Galileo, all combined some optics with astronomy. Tycho and the old ones combined alchemy; Newton dabbled in this also.)

Newton reached the age of twenty-three in 1665, the year of the Great Plague. The plague broke out in Cambridge as well as in London, and the whole college was sent down. Newton went back to Woolsthorpe, his mind teeming with ideas, and spent the rest of this year and part of the next in quiet pondering. Somehow or other he had got hold of the notion of centrifugal force. It was six years before Huyghens discovered and published the laws of centrifugal force, but in some quiet way of his own Newton knew about it and applied the idea to the motion of the planets.

We can almost follow the course of his thoughts as he brooded and meditated on the great problem which had taxed so many previous thinkers,—What makes the planets move round the sun? Kepler had discovered how they moved, but why did they so move, what urged them?

Even the "how" took a long time—all the time of the Greeks, through Ptolemy, the Arabs, Copernicus, Tycho: circular motion, epicycles, and excentrics had been the prevailing theory. Kepler, with his marvellous industry, had wrested from Tycho's observations the secret of their orbits. They moved in ellipses with the sun in one focus. Their rate of description of area, not their speed, was uniform and proportional to time.

Yes, and a third law, a mysterious law of unintelligible import, had also yielded itself to his penetrating industry—a law the discovery of which had given him the keenest delight, and excited an outburst of rapture—viz. that there was a relation between the distances and the periodic times of the several planets. The cubes of the distances were proportional to the squares of the times for the whole system. This law, first found true for the six primary planets, he had also extended, after Galileo's discovery, to the four secondary planets, or satellites of Jupiter (p. 81).

But all this was working in the dark—it was only the first step—this empirical discovery of facts; the facts were so, but how came they so? What made the planets move in this particular way? Descartes's vortices was an attempt, a poor and imperfect attempt, at an explanation. It had been hailed and adopted throughout Europe for want of a better, but it did not satisfy Newton. No, it proceeded on a wrong tack, and Kepler had proceeded on a wrong tack in imagining spokes or rays sticking out from the sun and driving the planets round like a piece of mechanism or mill work. For, note that all these theories are based on a wrong idea—the idea, viz., that some force is necessary to maintain a body in motion. But this was contrary to the laws of motion as discovered by Galileo. You know that during his last years of blind helplessness at Arcetri, Galileo had pondered and written much on the laws of motion, the foundation of mechanics. In his early youth, at Pisa, he had been similarly occupied; he had discovered the pendulum, he had refuted the Aristotelians by dropping weights from the leaning tower (which we must rejoice that no earthquake has yet injured), and he had returned to mechanics at intervals all his life; and now, when his eyes were useless for astronomy, when the outer world has become to him only a prison to be broken by death, he returns once more to the laws of motion, and produces the most solid and substantial work of his life.

For this is Galileo's main glory—not his brilliant exposition of the Copernican system, not his flashes of wit at the expense of a moribund philosophy, not his experiments on floating bodies, not even his telescope and astronomical discoveries—though these are the most taking and dazzling at first sight. No; his main glory and title to immortality consists in this, that he first laid the foundation of mechanics on a firm and secure basis of experiment, reasoning, and observation. He first discovered the true Laws of Motion.

I said little of this achievement in my lecture on him; for the work was written towards the end of his life, and I had no time then. But I knew I should have to return to it before we came to Newton, and here we are.

You may wonder how the work got published when so many of his manuscripts were destroyed. Horrible to say, Galileo's own son destroyed a great bundle of his father's manuscripts, thinking, no doubt, thereby to save his own soul. This book on mechanics was not burnt, however. The fact is it was rescued by one or other of his pupils, Toricelli or Viviani, who were allowed to visit him in his last two or three years; it was kept by them for some time, and then published surreptitiously in Holland. Not that there is anything in it bearing in any visible way on any theological controversy; but it is unlikely that the Inquisition would have suffered it to pass notwithstanding.

I have appended to the summary preceding this lecture (p. 160) the three axioms or laws of motion discovered by Galileo. They are stated by Newton with unexampled clearness and accuracy, and are hence known as Newton's laws, but they are based on Galileo's work. The first is the simplest; though ignorance of it gave the ancients a deal of trouble. It is simply a statement that force is needed to change the motion of a body; *i.e.* that if no force act on a body it will continue to move uniformly both in speed and direction—in other words, steadily, in a straight line. The old idea had been that some force was needed to maintain motion. On the contrary, the first law asserts, some force is needed to destroy it. Leave a body alone, free from all friction or other retarding forces, and it will go on for ever. The planetary motion through empty space therefore wants no keeping up; it is not the motion that demands a force to maintain it, it is the curvature of the path that needs a force to produce it continually. The motion of a planet is approximately uniform so far as speed is concerned, but it is not constant in direction; it is nearly a circle. The real force needed is not a propelling but a deflecting force.

The second law asserts that when a force acts, the motion changes, either in speed or in direction, or both, at a pace proportional to the magnitude of the force, and in the same direction as that in which the force acts. Now since it is almost solely in direction that planetary motion alters, a deflecting force only is needed; a force at right angles to the direction of motion, a force normal to the path. Considering the motion as circular, a force along the radius, a radial or centripetal force, must be acting continually. Whirl a weight round and round by a bit of elastic, the elastic is stretched; whirl it faster, it is stretched more. The moving mass pulls at the elastic—that is its centrifugal force; the hand at the centre pulls also—that is centripetal force.

The third law asserts that these two forces are equal, and together constitute the tension in the elastic. It is impossible to have one force alone, there must be a pair. You can't push hard against a body that offers no resistance. Whatever force you exert upon a body, with that same force the body must react upon you. Action and reaction are always equal and opposite.

Sometimes an absurd difficulty is felt with respect to this, even by engineers. They say, "If the cart pulls against the horse with precisely the same force as the horse pulls the cart, why should the cart move?" Why on earth not? The cart moves because the horse pulls it, and because nothing else is pulling it back. "Yes," they say, "the cart is pulling back." But what is it pulling back? Not itself, surely? "No, the horse." Yes, certainly the cart is pulling at the horse; if the cart offered no resistance what would be the good of the horse? That is what he is for, to overcome the pull-back of the cart; but nothing is pulling the cart back (except, of course, a little friction), and the horse is pulling it forward, hence it goes forward. There is no puzzle at all when once you realise that there are two bodies and two forces acting, and that one force acts on each body.[16]

If, indeed, two balanced forces acted on one body that would be in equilibrium, but the two equal forces contemplated in the third law act on two different bodies, and neither is in equilibrium.

So much for the third law, which is extremely simple, though it has extraordinarily far-reaching consequences, and when combined with a denial of "action at a distance," is precisely the principle of the Conservation of Energy. Attempts at perpetual motion may all be regarded as attempts to get round this "third law."

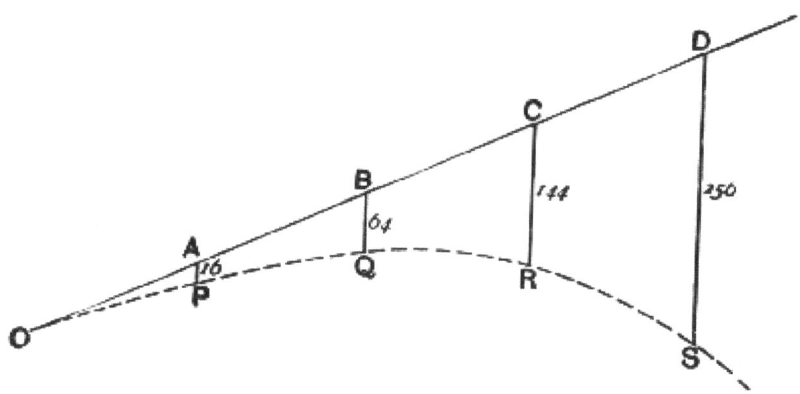

FIG. 57.

On the subject of the *second* law a great deal more has to be said before it can be in any proper sense even partially appreciated, but a complete discussion of it would involve a treatise on mechanics. It is *the* law of mechanics. One aspect of it we must attend to now in order to deal with the motion of the planets, and that is the fact that the change of motion of a body depends solely and simply on the force acting, and not at all upon what the body happens to be doing at the time it acts. It may be stationary, or it may be moving in any direction; that makes no difference.

Thus, referring back to the summary preceding Lecture IV, it is there stated that a dropped body falls 16 feet in the first second, that in two seconds it falls 64 feet, and so on, in proportion to the square of the time. So also will it be the case with a thrown body, but the drop must be reckoned from its line of motion—the straight line which, but for gravity, it would describe.

Thus a stone thrown from *O* with the velocity *OA* would in one second find itself at *A*, in two seconds at *B*, in three seconds at *C*, and so on, in accordance with the first law of motion, if no force acted. But if gravity acts it will have fallen 16 feet by the time it would have got to *A*, and so will find itself at *P*. In two seconds it will be at *Q*, having fallen a vertical height of 64 feet; in three seconds it will be at *R*, 144 feet below *C*; and so on. Its actual path will be a curve, which in this case is a parabola. (Fig. 57.)

If a cannon is pointed horizontally over a level plain, the cannon ball will be just as much affected by gravity as if it were dropped, and so will strike the plain at the same instant as another which was simply dropped where it started. One ball may have gone a mile and the other only dropped a hundred feet or so, but the time needed by both for the vertical drop will be the same. The horizontal motion of one is an extra, and is due to the powder.

As a matter of fact the path of a projectile in vacuo is only approximately a parabola. It is instructive to remember that it is really an ellipse with one focus very distant, but not at infinity. One of its foci is the centre of the earth. A projectile is really a minute satellite of the earth's, and in vacuo it accurately obeys all Kepler's laws. It happens not to be able to complete its orbit, because it was started inconveniently close to the earth, whose bulk gets in its way; but in that respect the earth is to be reckoned as a gratuitous obstruction, like a target, but a target that differs from most targets in being hard to miss.

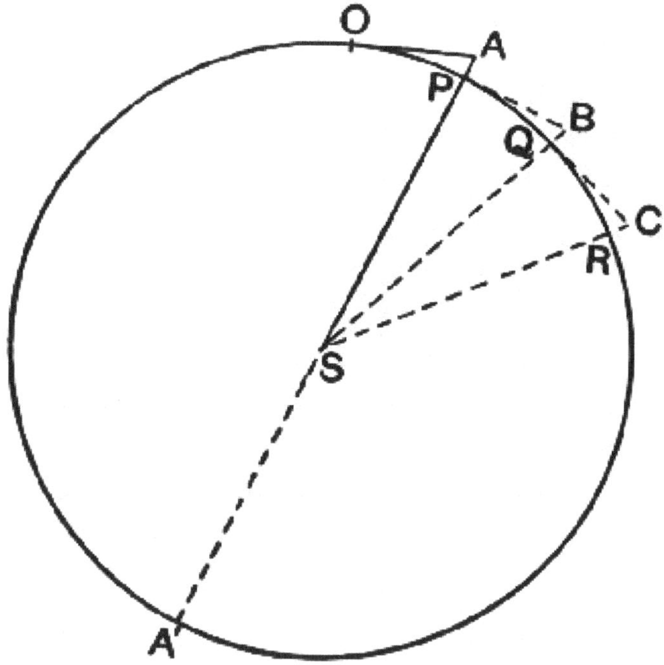

FIG. 58.

Now consider circular motion in the same way, say a ball whirled round by a string. (Fig. 58.)

Attending to the body at *O*, it is for an instant moving towards *A*, and if no force acted it would get to *A* in a time which for brevity we may call a second. But a force, the pull of the string, is continually drawing it towards *S*, and so it really finds itself at *P*, having described the circular arc *OP*, which may be considered to be compounded of, and analyzable into the rectilinear motion *OA* and the drop *AP*. At *P* it is for an instant moving towards *B*, and the same process therefore carries it to *Q*; in the third second it gets to *R*; and so on: always falling, so to speak, from its natural rectilinear path, towards the centre, but never getting any nearer to the centre.

The force with which it has thus to be constantly pulled in towards the centre, or, which is the same thing, the force with which it is tugging at whatever constraint it is that holds it in, is mv^2/r; where m is the mass of the particle, v its velocity, and r the radius of its circle of movement. This is the formula first given by Huyghens for centrifugal force.

We shall find it convenient to express it in terms of the time of one revolution, say T. It is easily done, since plainly T = circumference/speed $= 2\pi r/v$; so the above expression for centrifugal force becomes $4\pi^2 mr/T^2$.

As to the fall of the body towards the centre every microscopic unit of time, it is easily reckoned. For by Euclid III. 36, and Fig. 58, $AP.AA' = AO^2$. Take A very near O, then $OA = vt$, and $AA' = 2r$; so $AP = v^2t^2/2r = 2\pi^2 r\, t^2/T^2$; or the fall per second is $2\pi^2 r/T^2$, r being its distance from the centre, and T its time of going once round.

In the case of the moon for instance, r is 60 earth radii; more exactly 60·2; and T is a lunar month, or more precisely 27 days, 7 hours, 43 minutes, and 11½ seconds. Hence the moon's deflection from the tangential or rectilinear path every minute comes out as very closely 16 feet (the true size of the earth being used).

Returning now to the case of a small body revolving round a big one, and assuming a force directly proportional to the mass of both bodies, and inversely proportional to the square of the distance between them: *i.e.* assuming the known force of gravity, it is

$$\frac{V\,Mm}{r^2}$$

where V is a constant, called the gravitation constant, to be determined by experiment.

If this is the centripetal force pulling a planet or satellite in, it must be equal to the centrifugal force of this latter, viz. (see above).

$$\frac{4\pi^2 mr}{T^2}$$

Equate the two together, and at once we get

$$\frac{r^3}{T^2} = \frac{V}{4\pi^2}\mathrm{M};$$

or, in words, the cube of the distance divided by the square of the periodic time for every planet or satellite of the system under consideration, will be constant and proportional to the mass of the central body.

This is Kepler's third law, with a notable addition. It is stated above for circular motion only, so as to avoid geometrical difficulties, but even so it is very instructive. The reason of the proportion between r^3 and T^2 is at once manifest; and as soon as the constant V became known, *the mass of the central body*, the sun in the case of a planet, the earth in the case of the moon, Jupiter in the case of his satellites, was at once determined.

Newton's reasoning at this time might, however, be better displayed perhaps by altering the order of the steps a little, as thus:—

The centrifugal force of a body is proportional to r^3/T^2, but by Kepler's third law r^3/T^2 is constant for all the planets, reckoning r from the sun. Hence the centripetal force needed to hold in all the planets will be a single force emanating from the sun and varying inversely with the square of the distance from that body.

Such a force is at once necessary and sufficient. Such a force would explain the motion of the planets.

But then all this proceeds on a wrong assumption—that the planetary motion is circular. Will it hold for elliptic orbits? Will an inverse square law of force keep a body moving in an elliptic orbit about the sun in one focus? This is a far more difficult question. Newton solved it, but I do not believe that even he could have solved it, except that he had at his disposal two mathematical engines of great power—the Cartesian method of treating geometry, and his own method of Fluxions. One can explain the elliptic motion now mathematically, but hardly otherwise; and I must be content to state that the double fact is true—viz., that an inverse square law will move the body in an ellipse or other conic section with the sun in one focus, and that if a body so moves it *must* be acted on by an inverse square law.

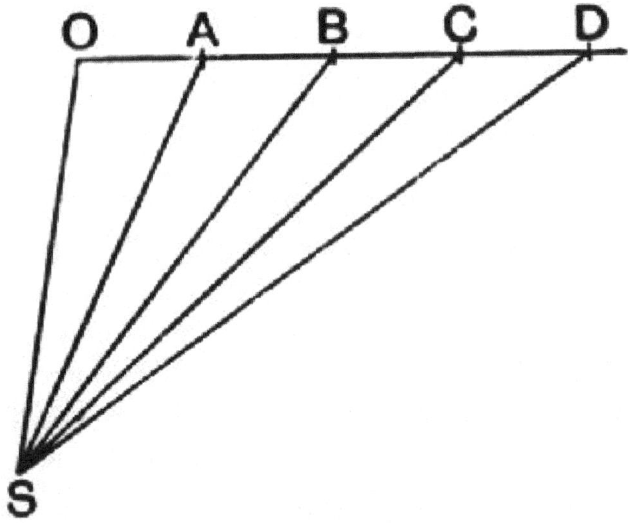

FIG. 59.

This then is the meaning of the first and third laws of Kepler. What about the second? What is the meaning of the equable description of areas? Well, that rigorously proves that a planet is acted on by a force directed to the centre about which the rate of description of areas is equable. It proves, in fact, that the sun is the attracting body, and that no other force acts.

For first of all if the first law of motion is obeyed, *i.e.* if no force acts, and if the path be equally subdivided to represent equal times, and straight lines be drawn from the divisions to any point whatever, all these areas thus enclosed will be equal, because they are triangles on equal base and of the same height (Euclid, I). See Fig. 59; *S* being any point whatever, and *A*, *B*, *C*, successive positions of a body.

Now at each of the successive instants let the body receive a sudden blow in the direction of that same point *S*, sufficient to carry it from *A* to *D* in the same time as it would have got to *B* if left alone. The result will be that there will be a compromise, and it will really arrive at *P*, travelling along the diagonal of the parallelogram *AP*. The area its radius vector sweeps out is therefore *SAP*, instead of what it would have been, *SAB*. But then these two areas are equal, because they are triangles on the same base *AS*, and between the same parallels *BP*, *AS*; for by the parallelogram law *BP* is parallel to *AD*. Hence the area that would have been described is described, and as all the areas were equal in the case of no force, they remain equal when the body receives a blow at the end of every equal interval of time, *provided* that every blow is actually directed to *S*, the point to which radii vectores are drawn.

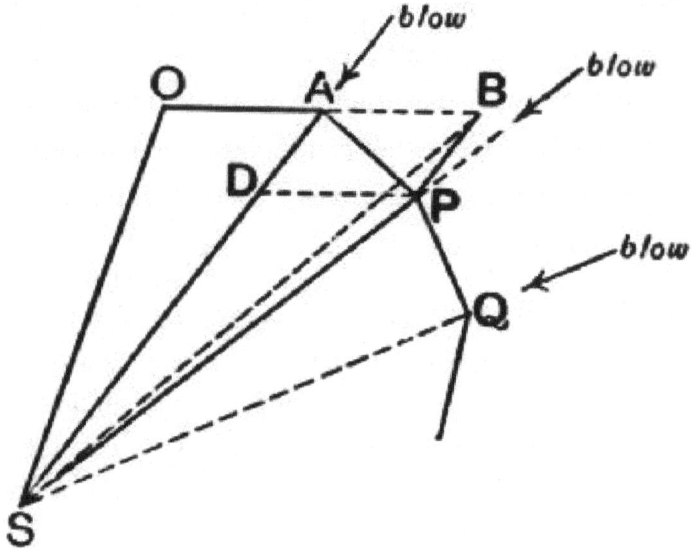

FIG. 60.

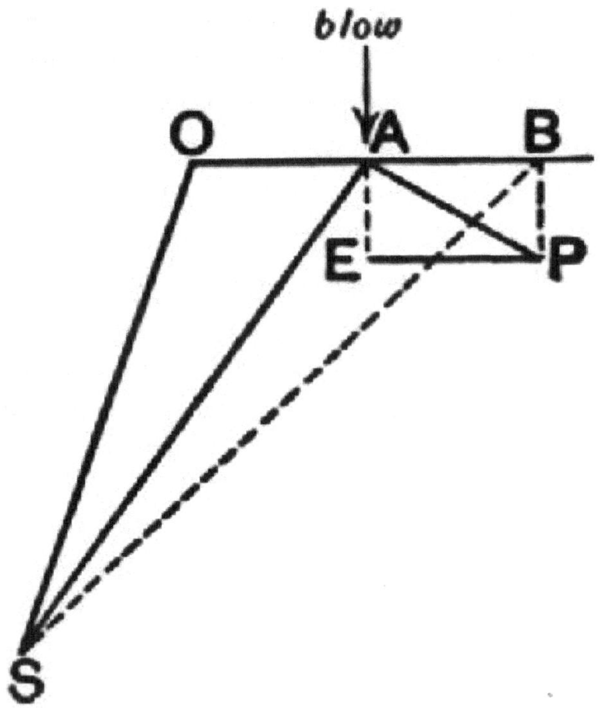

FIG. 61.

It is instructive to see that it does not hold if the blow is any otherwise directed; for instance, as in Fig. 61, when the blow is along *AE*, the body finds itself at *P* at the end of the second interval, but the area *SAP* is by no means equal to *SAB*, and therefore not equal to *SOA*, the area swept out in the first interval.

In order to modify Fig. 60 so as to represent continuous motion and steady forces, we have to take the sides of the polygon *OAPQ*, &c., very numerous and very small; in the limit, infinitely numerous and infinitely small. The path then becomes a curve, and the series of blows becomes a steady force directed towards *S*. About whatever point therefore the rate of description of areas is uniform, that point and no other must be the centre of all the force there is. If there be no force, as in Fig. 59, well and good, but if there be any force however small not directed towards *S*, then the rate of description of areas about *S* cannot be uniform. Kepler, however, says that the rate of description of areas of each planet about the sun is, by Tycho's observations, uniform; hence the sun is the centre of all the force that acts on them, and there is no other force, not even friction. That is the moral of Kepler's second law.

We may also see from it that gravity does not travel like light, so as to take time on its journey from sun to planet; for, if it did, there would be a sort of aberration, and the force on its arrival could no longer be accurately directed to the centre of the sun. (See *Nature*, vol. xlvi., p. 497.) It is a matter for accuracy of observation, therefore, to decide whether the minutest trace of such deviation can be detected, *i.e.* within what limits of accuracy Kepler's second law is now known to be obeyed.

I will content myself by saying that the limits are extremely narrow. [Reference may be made also to p. 208.]

Thus then it became clear to Newton that the whole solar system depended on a central force emanating from the sun, and varying inversely with the square of the distance from him: for by that hypothesis all the laws of Kepler concerning these motions were completely accounted for; and, in fact, the laws necessitated the hypothesis and established it as a theory.

Similarly the satellites of Jupiter were controlled by a force emanating from Jupiter and varying according to the same law. And again our moon must be controlled by a force from the earth, decreasing with the distance according to the same law.

Grant this hypothetical attracting force pulling the planets towards the sun, pulling the moon towards the earth, and the whole mechanism of the solar system is beautifully explained.

If only one could be sure there was such a force! It was one thing to calculate out what the effects of such a force would be: it was another to be able to put one's finger upon it and say, this is the force that actually exists and is known to exist. We must picture him meditating in his garden on this want—an attractive force towards the earth.

If only such an attractive force pulling down bodies to the earth existed. An apple falls from a tree. Why, it does exist! There is gravitation, common gravity that makes bodies fall and gives them their weight.

Wanted, a force tending towards the centre of the earth. It is to hand!

It is common old gravity that had been known so long, that was perfectly familiar to Galileo, and probably to Archimedes. Gravity that regulates the motion of projectiles. Why should it only pull stones and apples? Why should it not reach as high as the moon? Why should it not be the gravitation of the sun that is the central force acting on all the planets?

Surely the secret of the universe is discovered! But, wait a bit; is it discovered? Is this force of gravity sufficient for the purpose? It must vary inversely with the square of the distance from the centre of the earth. How far is the moon away? Sixty earth's radii. Hence the force of gravity at the moon's distance can only be 1⁄3600 of what it is on the earth's surface. So, instead of pulling it 16 ft. per second, it should pull it 16⁄3600 ft. per second, or 16 ft. a minute.[17] How can one decide whether such a force is able to pull the moon the actual amount required? To Newton this would seem only like a sum in arithmetic. Out with a pencil and paper and reckon how much the moon falls toward the earth in every second of its motion. Is it 16⁄3600? That is what it ought to be: but is it? The size of the earth comes into the calculation. Sixty miles make a degree, 360 degrees a circumference. This gives as the earth's diameter 6,873 miles; work it out.

The answer is not 16 feet a minute, it is 13·9 feet.

Surely a mistake of calculation?

No, it is no mistake: there is something wrong in the theory, gravity is too strong.

Instead of falling toward the earth 5⅓ hundredths of an inch every second, as it would under gravity, the moon only falls 4⅔ hundredths of an inch per second.

With such a discovery in his grasp at the age of twenty-three he is disappointed—the figures do not agree, and he cannot make them agree. Either gravity is not the force in action, or else something interferes with it. Possibly, gravity does part of the work, and the vortices of Descartes interfere with it.

He must abandon the fascinating idea for the time. In his own words, "he laid aside at that time any further thought of the matter."

So far as is known, he never mentioned his disappointment to a soul. He might, perhaps, if he had been at Cambridge, but he was a shy and solitary youth, and just as likely he might not. Up in Lincolnshire, in the seventeenth century, who was there for him to consult?

True, he might have rushed into premature publication, after our nineteenth century fashion, but that was not his method. Publication never seemed to have occurred to him.

His reticence now is noteworthy, but later on it is perfectly astonishing. He is so absorbed in making discoveries that he actually has to be reminded to tell any one about them, and some one else always has to see to the printing and publishing for him.

I have entered thus fully into what I conjecture to be the stages of this early discovery of the law of gravitation, as applicable to the heavenly bodies, because it is frequently and commonly misunderstood. It is sometimes thought that he discovered the force of gravity; I hope I have made it clear that he did no such thing. Every educated man long before his time, if asked why bodies fell, would reply just as glibly as they do now, "Because the earth attracts them," or "because of the force of gravity."

His discovery was that the motions of the solar system were due to the action of a central force, directed to the body at the centre of the system, and varying inversely with the square of the distance from it. This discovery was based upon Kepler's laws, and was clear and certain. It might have been published had he so chosen.

But he did not like hypothetical and unknown forces; he tried to see whether the known force of gravity would serve. This discovery at that time he failed to make, owing to a wrong numerical datum. The size of the earth he only knew from the common doctrine of sailors that 60 miles make a degree; and that threw him out. Instead of falling 16 feet a minute, as it ought under gravity, it only fell 13·9 feet, so he abandoned the idea. We do not find that he returned to it for sixteen years.

LECTURE VIII

NEWTON AND THE LAW OF GRAVITATION

WE left Newton at the age of twenty-three on the verge of discovering the mechanism of the solar system, deterred therefrom only by an error in the then imagined size of the earth. He had proved from Kepler's laws that a centripetal force directed to the sun, and varying as the inverse square of the distance from that body, would account for the observed planetary motions, and that a similar force directed to the earth would account for the lunar motion; and it had struck him that this force might be the very same as the familiar force of gravitation which gave to bodies their weight: but in attempting a numerical verification of this idea in the case of the moon he was led by the then received notion that sixty miles made a degree on the earth's surface into an erroneous estimate of the size of the moon's orbit. Being thus baffled in obtaining such verification, he laid the matter aside for a time.

The anecdote of the apple we learn from Voltaire, who had it from Newton's favourite niece, who with her husband lived and kept house for him all his later life. It is very like one of those anecdotes which are easily invented and believed in, and very often turn out on scrutiny to have no foundation. Fortunately this anecdote is well authenticated, and moreover is intrinsically probable; I say fortunately, because it is always painful to have to give up these child-learnt anecdotes, like Alfred and the cakes and so on. This anecdote of the apple we need not resign. The tree was blown down in 1820 and part of its wood is preserved.

I have mentioned Voltaire in connection with Newton's philosophy. This acute critic at a later stage did a good deal to popularise it throughout Europe and to overturn that of his own countryman Descartes. Cambridge rapidly became Newtonian, but Oxford remained Cartesian for fifty years or more. It is curious what little hold science and mathematics have ever secured in the older and more ecclesiastical University. The pride of possessing Newton has however no doubt been the main stimulus to the special pursuits of Cambridge.

He now began to turn his attention to optics, and, as was usual with him, his whole mind became absorbed in this subject as if nothing else had ever occupied him. His cash-book for this time has been discovered, and the entries show that he is buying prisms and lenses and polishing powder at the beginning of 1667. He was anxious to improve telescopes by making more perfect lenses than had ever been used before. Accordingly he calculated out their proper curves, just as Descartes had also done, and then proceeded to grind them as near as he could to those figures. But the images did not please him; they were always blurred and rather indistinct.

At length, it struck him that perhaps it was not the lenses but the light which was at fault. Perhaps light was so composed that it *could* not be focused accurately to a sharp and definite point. Perhaps the law of refraction was not quite accurate, but only an approximation. So he bought a prism to try the law. He let in sunlight through a small round hole in a window shutter, inserted the prism in the light, and received the deflected beam on a white screen; turning the prism about till it was deviated as little as possible. The patch on the screen was not a round disk, as it would have been without the prism, but was an elongated oval and was coloured at its extremities. Evidently refraction was not a simple geometrical deflection of a ray, there was a spreading out as well.

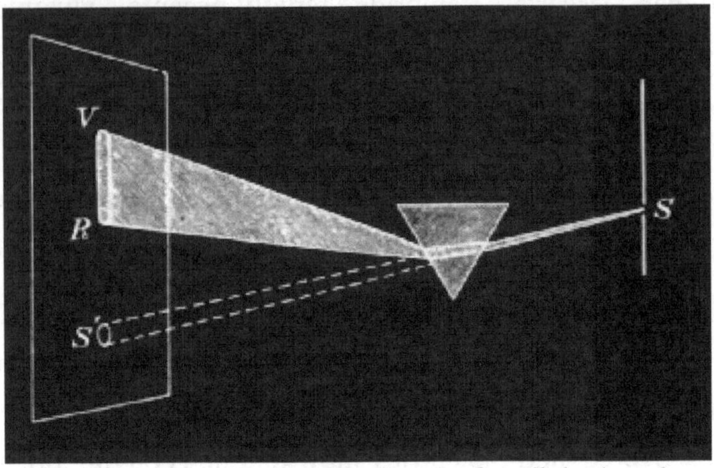

FIG. 63.—A prism not only *deviates* a beam of sunlight, but also spreads it out or *disperses* it.

Why did the image thus spread out? If it were due to irregularities in the glass a second prism should rather increase them, but a second prism when held in appropriate position was able to neutralise the dispersion and to reproduce the simple round white spot without deviation. Evidently the spreading out of the beam was connected in some definite way with its refraction. Could it be that the light particles after passing through the prism travelled in variously curved lines, as spinning racquet balls do? To examine this he measured the length of the oval patch when the screen was at different distances from the prism, and found that the two things were directly proportional to each other. Doubling the distance of the screen doubled the length of the patch. Hence the rays travelled in straight lines from the prism, and the spreading out was due to something that occurred within its substance. Could it be that white light was compound, was a mixture of several constituents, and that its different constituents were differently bent? No sooner thought than tried. Pierce the screen to let one of the constituents through and interpose a second prism in its path. If the spreading out depended on the prism only it should spread out just as much as before, but if it depended on the complex character of white light, this isolated simple constituent should be able to spread out no more. It did not spread out any more: a prism had no more dispersive power over it; it was deflected by the appropriate amount, but it was not analysed into constituents. It differed from sunlight in being simple. With many ingenious and beautifully simple experiments, which are quoted in full in several books on optics, he clinched the argument and established his discovery. White light was not simple but compound. It could be sorted out by a prism into an infinite number of constituent parts which were differently refracted, and the most striking of which Newton named violet, indigo, blue, green, yellow, orange, and red.

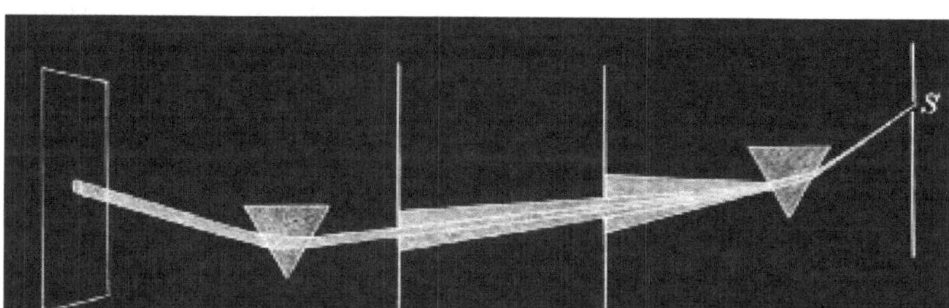

FIG. 64.—A single constituent of white light, obtained by the use of perforated screens is capable of no more dispersion.

At once the true nature of colour became manifest. Colour resided not in the coloured object as had till now been thought, but in the light which illuminated it. Red glass for instance adds nothing to sunlight. The light does not get dyed red by passing through the glass; all that the red glass does is to stop and absorb a large part of the sunlight; it is opaque to the larger portion, but it is transparent to that particular portion which affects our eyes with the sensation of red. The prism acts like a sieve sorting out the different kinds of light. Coloured media act like filters, stopping certain kinds but allowing the rest to go through. Leonardo's and all the ancient doctrines of colour had been singularly wrong; colour is not in the object but in the light.

Goethe, in his *Farbenlehre*, endeavoured to controvert Newton, and to reinstate something more like the old views; but his failure was complete.

Refraction analysed out the various constituents of white light and displayed them in the form of a series of overlapping images of the aperture, each of a different colour; this series of images we call a spectrum, and the operation we now call spectrum analysis. The reason of the defect of lenses was now plain: it was not so much a defect of the lens as a defect of light. A lens acts by refraction and brings rays to a focus. If light be simple it acts well, but if ordinary white light fall upon a lens, its different constituents have different foci; every bright object is fringed with colour, and nothing like a clear image can be obtained.

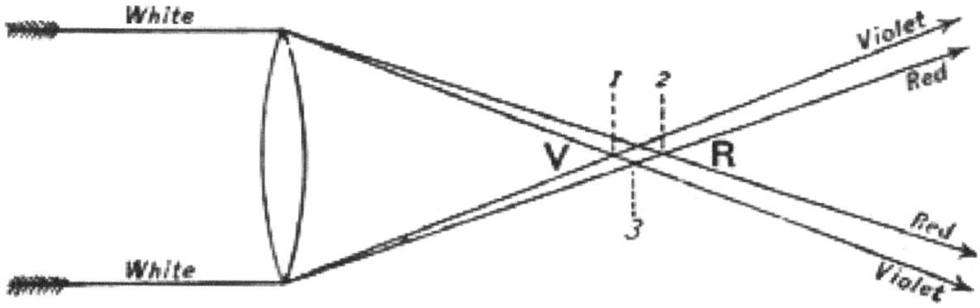

FIG. 65.—Showing the boundary rays of a parallel beam passing through a lens.

A parallel beam passing through a lens becomes conical; but instead of a single cone it is a sheaf or nest of cones, all having the edge of the lens as base, but each having a different vertex. The violet cone is innermost, near the lens, the red cone outermost, while the others lie between. Beyond the crossing point or focus the order of cones is reversed, as the above figure shows. Only the two marginal rays of the beam are depicted.

If a screen be held anywhere nearer the lens than the place marked 1 there will be a whitish centre to the patch of light and a red and orange fringe or border. Held anywhere beyond the region 2, the border of the patch will be blue and violet. Held about 3 the colour will be less marked than elsewhere, but nowhere can it be got rid of. Each point of an object will be represented in the image not by a point but by a coloured patch: a fact which amply explains the observed blurring and indistinctness.

Newton measured and calculated the distance between the violet and red foci—VR in the diagram—and showed that it was 1⁄50th the diameter of the lens. To overcome this difficulty (called chromatic aberration) telescope glasses were made small and of very long focus: some of them so long that they had no tube, all of them egregiously cumbrous. Yet it was with such instruments that all the early discoveries were made. With such an instrument, for instance, Huyghens discovered the real shape of Saturn's ring.

The defects of refractors seemed irremediable, being founded in the nature of light itself. So he gave up his "glass works"; and proceeded to think of reflexion from metal specula. A concave mirror forms an image just as a lens does, but since it does so without refraction or transmission through any substance, there is no accompanying dispersion or chromatic aberration.

The first reflecting telescope he made was 1 in. diameter and 6 in. long, and magnified forty times. It acted as well as a three or four feet refractor of that day, and showed Jupiter's moons. So he made a larger one, now in the library of the Royal Society, London, with an inscription:

"The first reflecting telescope, invented by Sir Isaac Newton, and made with his own hands."

This has been the parent of most of the gigantic telescopes of the present day. Fifty years elapsed before it was much improved on, and then, first by Hadley and afterwards by Herschel and others, large and good reflectors were constructed.

The largest telescope ever made, that of Lord Rosse, is a Newtonian reflector, fifty feet long, six feet diameter, with a mirror weighing four tons. The sextant, as used by navigators, was also invented by Newton.

The year after the plague, in 1667, Newton returned to Trinity College, and there continued his experiments on optics. It is specially to be noted that at this time, at the age of twenty-four, Newton had laid the foundations of all his greatest discoveries:—

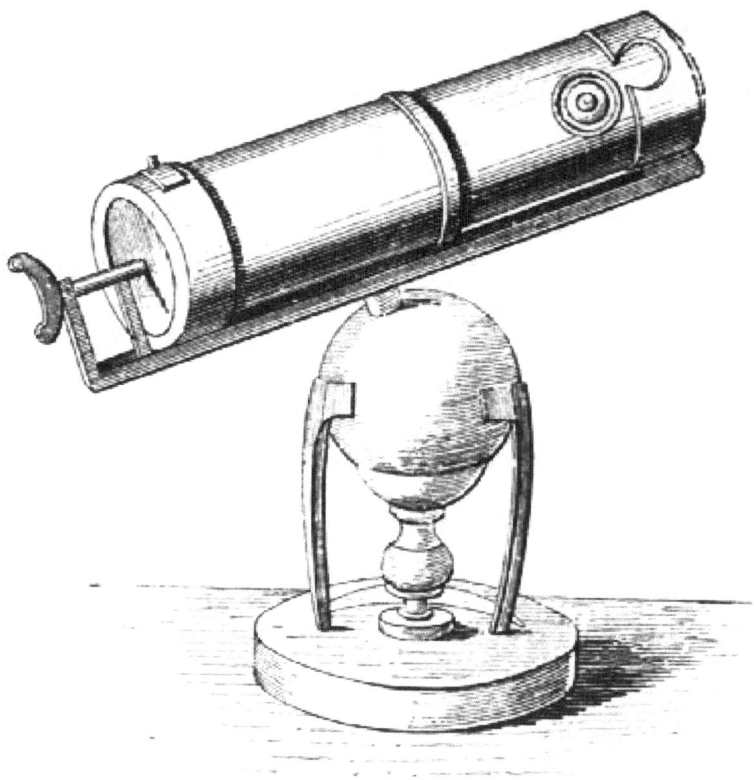

FIG. 66.—Newton's telescope.

The Theory of Fluxions; or, the Differential Calculus.

The Law of Gravitation; or, the complete theory of astronomy.

The compound nature of white light; or, the beginning of Spectrum Analysis.

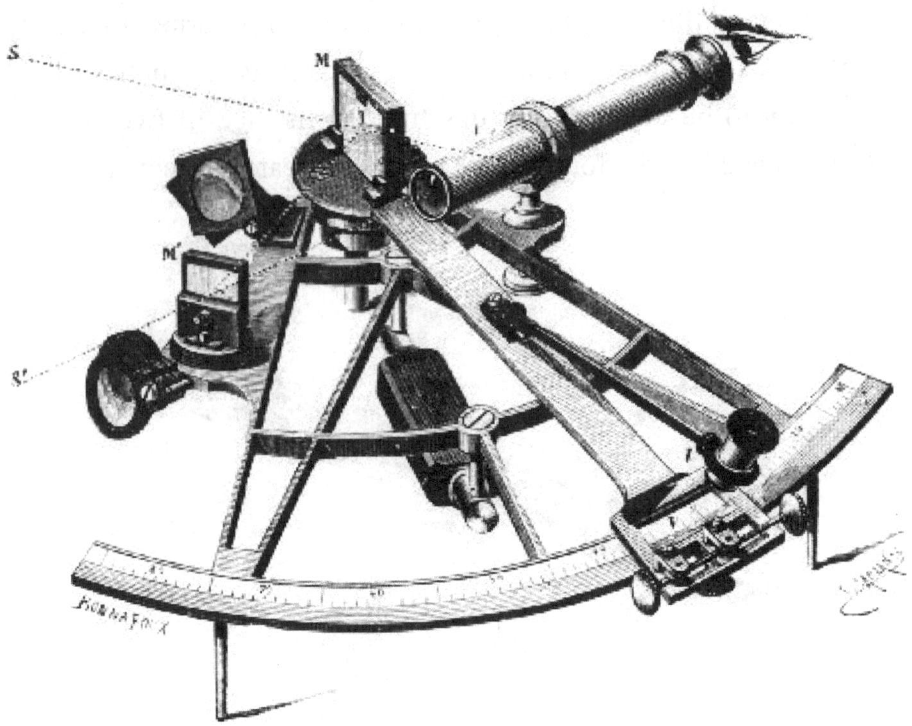

FIG. 67.—The sextant, as now made.

His later life was to be occupied in working these incipient discoveries out. But the most remarkable thing is that no one knew about any one of them. However, he was known as an accomplished young mathematician, and was made a fellow of his college. You remember that he had a friend there in the person of Dr. Isaac Barrow, first Lucasian Professor of Mathematics in the University. It happened, about 1669, that a mathematical discovery of some interest was being much discussed, and Dr. Barrow happened to mention it to Newton, who said yes, he had worked out that and a few other similar things some time ago. He accordingly went and fetched some papers to Dr. Barrow, who forwarded them to other distinguished mathematicians, and it thus appeared that Newton had discovered theorems much more general than this special case that was exciting so much interest. Dr. Barrow, being anxious to devote his time more particularly to theology, resigned his chair the same year in favour of Newton, who was accordingly elected to the Lucasian Professorship, which he held for thirty years. This chair is now the most famous in the University, and it is commonly referred to as the chair of Newton.

Still, however, his method of fluxions was unknown, and still he did not publish it. He lectured first on optics, giving an account of his experiments. His lectures were afterwards published both in Latin and English, and are highly valued to this day.

The fame of his mathematical genius came to the ears of the Royal Society, and a motion was made to get him elected a fellow of that body. The Royal Society, the oldest and most famous of all scientific societies with a continuous existence, took its origin in some private meetings, got up in London by the Hon. Robert Boyle and a few scientific friends, during all the trouble of the Commonwealth.

After the restoration, Charles II. in 1662 incorporated it under Royal Charter; among the original members being Boyle, Hooke, Christopher Wren, and other less famous names. Boyle was a great experimenter, a worthy follower of Dr. Gilbert. Hooke began as his assistant, but being of a most extraordinary ingenuity he rapidly rose so as to exceed his master in importance. Fate has been a little unkind to Hooke in placing him so near to Newton; had he lived in an ordinary age he would undoubtedly have shone as a star of the first magnitude. With great ingenuity, remarkable scientific insight, and consummate experimental skill, he stands in many respects almost on a level with Galileo. But it is difficult to see stars even of the first magnitude when the sun is up, and thus it happens that the name and fame of this brilliant man are almost lost in the blaze of Newton. Of Christopher Wren I need not say much. He is well known as an architect, but he was a most accomplished all-round man, and had a considerable taste and faculty for science.

These then were the luminaries of the Royal Society at the time we are speaking of, and to them Newton's first scientific publication was submitted. He communicated to them an account of his reflecting telescope, and presented them with the instrument.

Their reception of it surprised him; they were greatly delighted with it, and wrote specially thanking him for the communication, and assuring him that all right should be done him in the matter of the invention. The Bishop of Salisbury (Bishop Burnet) proposed him for election as a fellow, and elected he was.

In reply, he expressed his surprise at the value they set on the telescope, and offered, if they cared for it, to send them an account of a discovery which he doubts not will prove much more grateful than the communication of that instrument, "being in my judgment the oddest, if not the most considerable detection that has recently been made into the operations of Nature."

So he tells them about his optical researches and his discovery of the nature of white light, writing them a series of papers which were long afterwards incorporated and published as his *Optics*. A magnificent work, which of itself suffices to place its author in the first rank of the world's men of science.

The nature of white light, the true doctrine of colour, and the differential calculus! besides a good number of minor results—binomial theorem, reflecting telescope, sextant, and the like; one would think it enough for one man's life-work, but the masterpiece remains still to be mentioned. It is as when one is considering Shakspeare: *King Lear*, *Macbeth*, *Othello*,—surely a sufficient achievement,—but the masterpiece remains.

Comparisons in different departments are but little help perhaps, nevertheless it seems to me that in his own department, and considered simply as a man of science, Newton towers head and shoulders over, not only his contemporaries—that is a small matter—but over every other scientific man who has ever lived, in a way that we can find no parallel for in other departments. Other nations admit his scientific pre-eminence with as much alacrity as we do.

Well, we have arrived at the year 1672 and his election to the Royal Society. During the first year of his membership there was read at one of the meetings a paper giving an account of a very careful determination of the length of a degree (*i.e.* of the size of the earth), which had been made by Picard near Paris. The length of the degree turned out to be not sixty miles, but nearly seventy miles. How soon Newton heard of this we do not learn—probably not for some years,—Cambridge was not so near London then as it is now, but ultimately it was brought to his notice. Armed with this new datum, his old speculation concerning gravity occurred to him. He had worked out the mechanics of the solar system on a certain hypothesis, but it had remained a hypothesis somewhat out of harmony with apparent fact. What if it should turn out to be true after all!

He took out his old papers and began again the calculation. If gravity were the force keeping the moon in its orbit, it would fall toward the earth sixteen feet every minute. How far did it fall? The newly known size of the earth would modify the figures: with intense excitement he runs through the working, his mind leaps before his hand, and as he perceives the answer to be coming out right, all the infinite meaning and scope of his mighty discovery flashes upon him, and he can no longer see the paper. He throws down the pen; and the secret of the universe is, to one man, known.

But of course it had to be worked out. The meaning might flash upon him, but its full detail required years of elaboration; and deeper and deeper consequences revealed themselves to him as he proceeded.

For two years he devoted himself solely to this one object. During those years he lived but to calculate and think, and the most ludicrous stories are told concerning his entire absorption and inattention to ordinary affairs of life. Thus, for instance, when getting up in a morning he would sit on the side of the bed half-dressed, and remain like that till dinner time. Often he would stay at home for days together, eating what was taken to him, but without apparently noticing what he was doing.

One day an intimate friend, Dr. Stukely, called on him and found on the table a cover laid for his solitary dinner. After waiting a long time, Dr. Stukely removed the cover and ate the chicken underneath it, replacing and covering up the bones again. At length Newton appeared, and after greeting his friend, sat down to dinner, but on lifting the cover he said in surprise, "Dear me, I thought I had not dined, but I see I have."

It was by this continuous application that the *Principia* was accomplished. Probably nothing of the first magnitude can be accomplished without something of the same absorbed unconsciousness and freedom from interruption. But though desirable and essential for the *work*, it was a severe tax upon the powers of the *man*. There is, in fact, no doubt that Newton's brain suffered temporary aberration after this effort for a short time. The attack was slight, and it has been denied; but there are letters extant which are inexplicable otherwise, and moreover after a year or two he writes to his friends apologizing for strange and disjointed epistles, which he believed he had written without understanding clearly what he wrote. The derangement was, however, both slight and temporary: and it is only instructive to us as showing at what cost such a work as the *Principia* must be produced, even by so mighty a mind as that of Newton.

The first part of the work having been done, any ordinary mortal would have proceeded to publish it; but the fact is that after he had sent to the Royal Society his papers on optics, there had arisen controversies and objections; most of them rather paltry, to which he felt compelled to find answers. Many men would have enjoyed this part of the work, and taken it as evidence of interest and success. But to Newton's shy and retiring disposition these discussions were merely painful. He writes, indeed, his answers with great patience and ability, and ultimately converts the more reasonable of his opponents, but he relieves his mind in the following letter to the secretary of the Royal Society: "I see I have made myself a slave to philosophy, but if I get free of this present business I will resolutely bid adieu to it eternally, except what I do for my private satisfaction or leave to come out after me; for I see a man must either resolve to put out nothing new, or to become a slave to defend it." And again in a letter to Leibnitz: "I have been so persecuted with discussions arising out of my theory of light that I blamed my own imprudence for parting with so substantial a blessing as my quiet to run after a shadow." This shows how much he cared for contemporary fame.

So he locked up the first part of the *Principia* in his desk, doubtless intending it to be published after his death. But fortunately this was not so to be.

In 1683, among the leading lights of the Royal Society, the same sort of notions about gravity and the solar system began independently to be bruited. The theory of gravitation seemed to be in the air, and Wren, Hooke, and Halley had many a talk about it.

Hooke showed an experiment with a pendulum, which he likened to a planet going round the sun. The analogy is more superficial than real. It does not obey Kepler's laws; still it was a striking experiment. They had guessed at a law of inverse squares, and their difficulty was to prove what curve a body subject to it would describe. They knew it ought to be an ellipse if it was to serve to explain the planetary motion, and Hooke said he could prove that an ellipse it was; but he was nothing of a mathematician, and the others scarcely believed him. Undoubtedly he had shrewd inklings of the truth, though his guesses were based on little else than a most sagacious intuition. He surmised also that gravity was the force concerned, and asserted that the path of an ordinary projectile was an ellipse, like the path of a planet—which is quite right. In fact the beginnings of the discovery were beginning to dawn upon him in the well-known way in which things do dawn upon ordinary men of genius: and had Newton not lived we should doubtless, by the labours of a long chain of distinguished men, beginning with Hooke, Wren, and Halley, have been now in possession of all the truths revealed by the *Principia*. We should never have had them stated in the same form, nor proved with the same marvellous lucidity and simplicity, but the facts themselves we should by this time have arrived at. Their developments and completions, due to such men as Clairaut, Euler, D'Alembert, Lagrange, Laplace, Airy, Leverrier, Adams, we should of course not have had to the same extent; because the lives and energies of these great men would have been partially consumed in obtaining the main facts themselves.

The youngest of the three questioners at the time we are speaking of was Edmund Halley, an able and remarkable man. He had been at Cambridge, doubtless had heard Newton lecture, and had acquired a great veneration for him.

In January, 1684, we find Wren offering Hooke and Halley a prize, in the shape of a book worth forty shillings, if they would either of them bring him within two months a demonstration that the path of a planet subject to an inverse square law would be an ellipse. Not in two months, nor yet in seven, was there any proof forthcoming. So at last, in August, Halley went over to Cambridge to speak to Newton about the difficult problem and secure his aid.

Arriving at his rooms he went straight to the point. He said, "What path will a body describe if it be attracted by a centre with a force varying as the inverse square of the distance." To which Newton at once replied, "An ellipse." "How on earth do you know?" said Halley in amazement. "Why, I have calculated it," and began hunting about for the paper. He actually couldn't find it just then, but sent it him shortly by post, and with it much more—in fact, what appeared to be a complete treatise on motion in general.

With his valuable burden Halley hastened to the Royal Society and told them what he had discovered. The Society at his representation wrote to Mr. Newton asking leave that it might be printed. To this he consented; but the Royal Society wisely appointed Mr. Halley to see after him and jog his memory, in case he forgot about it. However, he set to work to polish it up and finish it, and added to it a great number of later developments and embellishments, especially the part concerning the lunar theory, which gave him a deal of trouble—and no wonder; for in the way he has put it there never was a man yet living who could have done the same thing. Mathematicians regard the achievement now as men might stare at the work of some demigod of a bygone age, wondering what manner of man this was, able to wield such ponderous implements with such apparent ease.

To Halley the world owes a great debt of gratitude—first, for discovering the *Principia*; second, for seeing it through the press; and third, for defraying the cost of its publication out of his own scanty purse. For though he ultimately suffered no pecuniary loss, rather the contrary, yet there was considerable risk in bringing out a book which not a dozen men living could at the time comprehend.

It is no small part of the merit of Halley that he recognized the transcendent value of the yet unfinished work, that he brought it to light, and assisted in its becoming understood to the best of his ability.

Though Halley afterwards became Astronomer-Royal, lived to the ripe old age of eighty-six, and made many striking observations, yet he would be the first to admit that nothing he ever did was at all comparable in importance with his discovery of the *Principia*; and he always used to regard his part in it with peculiar pride and pleasure.

And how was the *Principia* received? Considering the abstruse nature of its subject, it was received with great interest and enthusiasm. In less than twenty years the edition was sold out, and copies fetched large sums. We hear of poor students copying out the whole in manuscript in order to possess a copy—not by any means a bad thing to do, however many copies one may possess. The only useful way really to read a book like that is to pore over every sentence: it is no book to be skimmed.

While the *Principia* was preparing for the press a curious incident of contact between English history and the University occurred. It seems that James II., in his policy of Catholicising the country, ordered both Universities to elect certain priests to degrees without the ordinary oaths. Oxford had given way, and the Dean of Christ Church was a creature of James's choosing. Cambridge rebelled, and sent eight of its members, among them Mr. Newton, to plead their cause before the Court of High Commission. Judge Jeffreys presided over the Court, and threatened and bullied with his usual insolence. The Vice-Chancellor of Cambridge was deprived of office, the other deputies were silenced and ordered away. From the precincts of this court of justice Newton returned to Trinity College to complete the *Principia*.

By this time Newton was only forty-five years old, but his main work was done. His method of fluxions was still unpublished; his optics was published only imperfectly; a second edition of the *Principia*, with additions and improvements, had yet to appear; but fame had now come upon him, and with fame worries of all kinds.

By some fatality, principally no doubt because of the interest they excited, every discovery he published was the signal for an outburst of criticism and sometimes of attack. I shall not go into these matters: they are now trivial enough, but it is necessary to mention them, because to Newton they evidently loomed large and terrible, and occasioned him acute torment.

FIG. 68.—Newton when young.
(*From an engraving by B. Reading after Sir Peter Lely.*)

No sooner was the *Principia* put than Hooke put in his claims for priority. And indeed his claims were not altogether negligible; for vague ideas of the same sort had been floating in his comprehensive mind, and he doubtless felt indistinctly conscious of a great deal more than he could really state or prove.

By indiscreet friends these two great men were set somewhat at loggerheads, and worse might have happened had they not managed to come to close quarters, and correspond privately in a quite friendly manner, instead of acting through the mischievous medium of third parties. In the next edition Newton liberally recognizes the claims of both Hooke and Wren. However, he takes warning betimes of what he has to expect, and writes to Halley that he will only publish the first two books, those containing general theorems on motion. The third book—concerning the system of the world, *i.e.* the application to the solar system—he says "I now design to suppress. Philosophy is such an impertinently litigious lady that a man had as good be engaged in law-suits as have to do with her. I found it so formerly, and now I am no sooner come near her again but she gives me warning. The two books without the third will not so well bear the title 'Mathematical Principles of Natural Philosophy,' and therefore I had altered it to this, 'On the Free Motion of Two Bodies'; but on second thoughts I retain the former title: 'twill help the sale of the book—which I ought not to diminish now 'tis yours."

However, fortunately, Halley was able to prevail upon him to publish the third book also. It is, indeed, the most interesting and popular of the three, as it contains all the direct applications to astronomy of the truths established in the other two.

Some years later, when his method of fluxions was published, another and a worse controversy arose—this time with Leibnitz, who had also independently invented the differential calculus. It was not so well recognized then how frequently it happens that two men independently and unknowingly work at the very same thing at the same time. The history of science is now full of such instances; but then the friends of each accused the other of plagiarism.

I will not go into the controversy: it is painful and useless. It only served to embitter the later years of two great men, and it continued long after Newton's death—long after both their deaths. It can hardly be called ancient history even now.

But fame brought other and less unpleasant distractions than controversies. We are a curious, practical, and rather stupid people, and our one idea of honouring a man is to *vote* for him in some way or other; so they sent Newton to Parliament. He went, I believe, as a Whig, but it is not recorded that he spoke. It is, in fact, recorded that he was once expected to speak when on a Royal Commission about some question of chronometers, but that he would not. However, I dare say he made a good average member.

Then a little later it was realized that Newton was poor, that he still had to teach for his livelihood, and that though the Crown had continued his fellowship to him as Lucasian Professor without the necessity of taking orders, yet it was rather disgraceful that he should not be better off. So an appeal was made to the Government on his behalf, and Lord Halifax, who exerted himself strongly in the matter, succeeding to office on the accession of William III., was able to make him ultimately Master of the Mint, with a salary of some £1,200 a year. I believe he made rather a good Master, and turned out excellent coins: certainly he devoted his attention to his work there in a most exemplary manner.

But what a pitiful business it all is! Here is a man sent by Heaven to do certain things which no man else could do, and so long as he is comparatively unknown he does them; but so soon as he is found out, he is clapped into a routine office with a big salary: and there is, comparatively speaking, an end of him. It is not to be supposed that he had lost his power, for he frequently solved problems very quickly which had been given out by great Continental mathematicians as a challenge to the world.

We may ask why Newton allowed himself to be thus bandied about instead of settling himself down to the work in which he was so pre-eminently great. Well, I expect your truly great man never realizes how great he is, and seldom knows where his real strength lies. Certainly Newton did not know it. He several times talks of giving up philosophy altogether; and though he never really does it, and perhaps the feeling is one only born of some temporary overwork, yet he does not sacrifice everything else to it as he surely must had he been conscious of his own greatness. No; self-consciousness was the last thing that affected him. It is for a great man's contemporaries to discover him, to make much of him, and to put him in surroundings where he may flourish luxuriantly in his own heaven-intended way.

However, it is difficult for us to judge of these things. Perhaps if he had been maintained at the national expense to do that for which he was preternaturally fitted, he might have worn himself out prematurely; whereas by giving him routine work the scientific world got the benefit of his matured wisdom and experience. It was no small matter to the young Royal Society to be able to have him as their President for twenty-four years. His portrait has hung over the President's chair ever since, and there I suppose it will continue to hang until the Royal Society becomes extinct.

The events of his later life I shall pass over lightly. He lived a calm, benevolent life, universally respected and beloved. His silver-white hair when he removed his peruke was a venerable spectacle. A lock of it is still preserved, with many other relics, in the library of Trinity College. He died quietly, after a painful illness, at the ripe age of eighty-five. His body lay in state in the Jerusalem Chamber, and he was buried in Westminster Abbey, six peers bearing the pall. These things are to be mentioned to the credit of the time and the country; for after we have seen the calamitous spectacle of the way Tycho and Kepler and Galileo were treated by their ungrateful and unworthy countries, it is pleasant to reflect that England, with all its mistakes, yet recognized *her* great man when she received him, and honoured him with the best she knew how to give.

FIG. 69.—Sir Isaac Newton.

Concerning his character, one need only say that it was what one would expect and wish. It was characterized by a modest, calm, dignified simplicity. He lived frugally with his niece and her husband, Mr. Conduit, who succeeded him as Master of the Mint. He never married, nor apparently did he ever think of so doing. The idea, perhaps, did not naturally occur to him, any more than the idea of publishing his work did.

He was always a deeply religious man and a sincere Christian, though somewhat of the Arian or Unitarian persuasion—so, at least, it is asserted by orthodox divines who understand these matters. He studied theology more or less all his life, and towards the end was greatly interested in questions of Biblical criticism and chronology. By some ancient eclipse or other he altered the recognized system of dates a few hundred years; and his book on the prophecies of Daniel and the Revelation of St. John, wherein he identifies the beast with the Church of Rome in quite the orthodox way, is still by some admired.

But in all these matters it is probable that he was a merely ordinary man, with natural acumen and ability doubtless, but nothing in the least superhuman. In science, the impression he makes upon me is only expressible by the words inspired, superhuman.

And yet if one realizes his method of work, and the calm, uninterrupted flow of all his earlier life, perhaps his achievements become more intelligible. When asked how he made his discoveries, he replied: "By always thinking unto them. I keep the subject constantly before me, and wait till the first dawnings open slowly by little and little into a full and clear light." That is the way—quiet, steady, continuous thinking, uninterrupted and unharassed brooding. Much may be done under those conditions. Much ought to be sacrificed to obtain those conditions. All the best thinking work of the world has been thus done.[18] Buffon said: "Genius is patience." So says Newton: "If I have done the public any service this way, it is due to nothing but industry and patient thought." Genius patience? No, it is not quite that, or, rather, it is much more than that; but genius without patience is like fire without fuel—it will soon burn itself out.

NOTES FOR LECTURE IX

The *Principia* published 1687.
Newton died 1727.

The Law of Gravitation.—Every particle of matter attracts every other particle of matter with a force proportional to the mass of each and to the inverse square of the distance between them.

Some of Newton's Deductions.

1. Kepler's second law (equable description of areas) proves that each planet is acted on by a force directed towards the sun as a centre of force.

2. Kepler's first law proves that this central force diminishes in the same proportion as the square of the distance increases.

3. Kepler's third law proves that all the planets are acted on by the same kind of force; of an intensity depending on the mass of the sun.[19]

4. So by knowing the length of year and distance of any planet from the sun, the sun's mass can be calculated, in terms of that of the earth.

5. For the satellites, the force acting depends on the mass of *their* central body, a planet. Hence the mass of any planet possessing a satellite becomes known.

6. The force constraining the moon in her orbit is the same gravity as gives terrestrial bodies their weight and regulates the motion of projectiles. [Because, while a stone drops 16 feet in a second, the moon, which is 60 times as far from the centre of the earth, drops 16 feet in a minute.]

7. The moon is attracted not only by the earth, but by the sun also; hence its orbit is perturbed, and Newton calculated out the chief of these perturbations, viz.:—

(The equation of the centre, discovered by Hipparchus.)

(*a*) The evection, discovered by Hipparchus and Ptolemy.

(*b*) The variation, discovered by Tycho Brahé.

(*c*) The annual equation, discovered by Tycho Brahé.

(*d*) The retrogression of the nodes, then being observed at Greenwich by Flamsteed.

(*e*) The variation of inclination, then being observed at Greenwich by Flamsteed.

(*f*) The progression of the apses (with an error of one-half).

(*g*) The inequality of apogee, previously unknown.

(*h*) The inequality of nodes, previously unknown.

8. Each planet is attracted not only by the sun but by the other planets, hence their orbits are slightly affected by each other. Newton began the theory of planetary perturbations.

9. He recognized the comets as members of the solar system, obedient to the same law of gravity and moving in very elongated ellipses; so their return could be predicted (*e.g.* Halley's comet).

10. Applying the idea of centrifugal force to the earth considered as a rotating body, he perceived that it could not be a true sphere, and calculated its oblateness, obtaining 28 miles greater equatorial than polar diameter.

11. Conversely, from the observed shape of Jupiter, or any planet, the length of its day could be estimated.

12. The so-calculated shape of the earth, in combination with centrifugal force, causes the weight of bodies to vary with latitude; and Newton calculated the amount of this variation. 194 lbs. at pole balance 195 lbs. at equator.

13. A homogeneous sphere attracts as if its mass were concentrated at its centre. For any other figure, such as an oblate spheroid, this is not exactly true. A hollow concentric spherical shell exerts no force on small bodies inside it.

14. The earth's equatorial protuberance, being acted on by the attraction of the sun and moon, must disturb its axis of rotation in a calculated manner; and thus is produced the precession of the

equinoxes. [The attraction of the planets on the same protuberance causes a smaller and rather different kind of precession.]

15. The waters of the ocean are attracted towards the sun and moon on one side, and whirled a little further away than the solid earth on the other side: hence Newton explained all the main phenomena of the tides.

16. The sun's mass being known, he calculated the height of the solar tide.

17. From the observed heights of spring and neap tides he determined the lunar tide, and thence made an estimate of the mass of the moon.

REFERENCE TABLE OF NUMERICAL DATA.

	Masses in Solar System.	Height dropped by a stone in first second.	Length of Day or time of rotation.
Mercury	·065	7·0 feet	24 hours
Venus	·885	15·8 "	23½ "
Earth	1·000	16·1 "	24 "
Mars	·108	6·2 "	24½ "
Jupiter	300·8	45·0 "	10 "
Saturn	89·7	18·4 "	10½ "
The Sun	316000·	436·0 "	608 "
The Moon	about ·012	3·7 "	702 "

The mass of the earth, taken above as unity, is 6,000 trillion tons.

Observatories.—Uraniburg flourished from 1576 to 1597; the Observatory of Paris was founded in 1667; Greenwich Observatory in 1675.

Astronomers-Royal.—Flamsteed, Halley, Bradley, Bliss, Maskelyne, Pond, Airy, Christie.

LECTURE IX

NEWTON'S "PRINCIPIA"

THE law of gravitation, above enunciated, in conjunction with the laws of motion rehearsed at the end of the preliminary notes of Lecture VII., now supersedes the laws of Kepler and includes them as special cases. The more comprehensive law enables us to criticize Kepler's laws from a higher standpoint, to see how far they are exact and how far they are only approximations. They are, in fact, not precisely accurate, but the reason for every discrepancy now becomes abundantly clear, and can be worked out by the theory of gravitation.

We may treat Kepler's laws either as immediate consequences of the law of gravitation, or as the known facts upon which that law was founded. Historically, the latter is the more natural plan, and it is thus that they are treated in the first three statements of the above notes; but each proposition may be worked inversely, and we might state them thus:—

1. The fact that the force acting on each planet is directed to the sun, necessitates the equable description of areas.

2. The fact that the force varies as the inverse square of the distance, necessitates motion in an ellipse, or some other conic section, with the sun in one focus.

3. The fact that one attracting body acts on all the planets with an inverse square law, causes the cubes of their mean distances to be proportional to the squares of their periodic times.

Not only these but a multitude of other deductions follow rigorously from the simple datum that every particle of matter attracts every other particle with a force directly proportional to the mass of each and to the inverse square of their mutual distance. Those dealt with in the *Principia* are summarized above, and it will be convenient to run over them in order, with the object of giving some idea of the general meaning of each, without attempting anything too intricate to be readily intelligible.

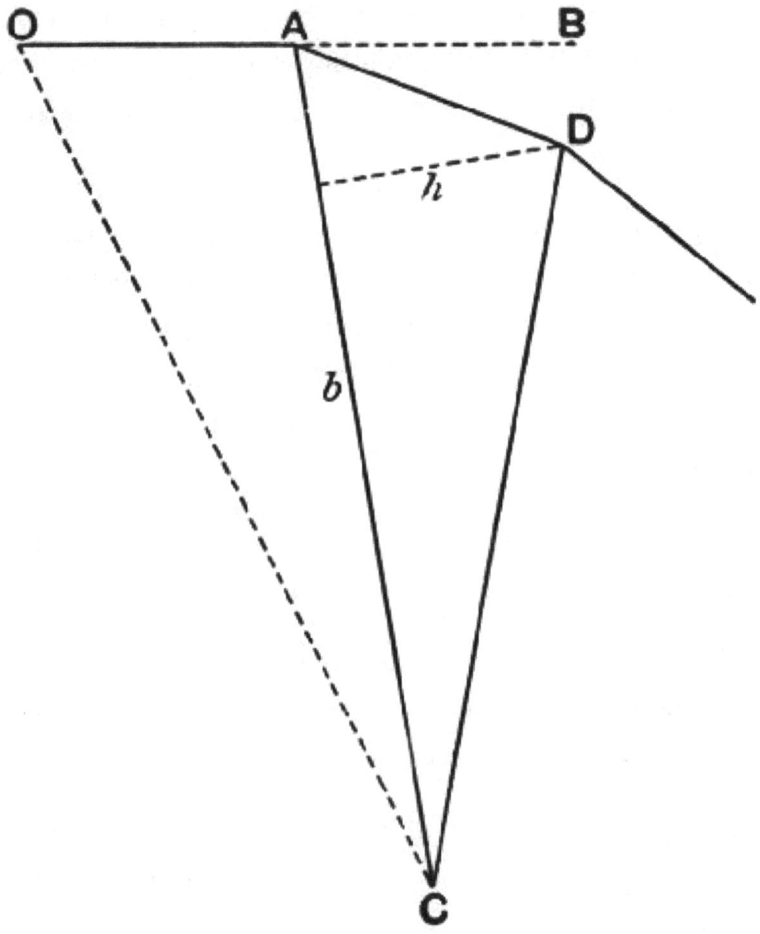

FIG. 70.

No. 1. Kepler's second law (equable description of areas) proves that each planet is acted on by a force directed towards the sun as a centre of force.

The equable description of areas about a centre of force has already been fully, though briefly, established. (p. 175.) It is undoubtedly of fundamental importance, and is the earliest instance of the serious discussion of central forces, *i.e.* of forces directed always to a fixed centre.

We may put it afresh thus:—OA has been the motion of a particle in a unit of time; at A it receives a knock towards C, whereby in the next unit it travels along AD instead of AB. Now the area of the triangle CAD, swept out by the radius vector in unit time, is $\frac{1}{2}bh$; h being the perpendicular height of the triangle from the base AC. (Fig. 70.) Now the blow at A, being along the base, has no effect upon h; and consequently the area remains just what it would have been without the blow. A blow directed to any point other than C would at once alter the area of the triangle.

One interesting deduction may at once be drawn. If gravity were a radiant force emitted from the sun with a velocity like that of light, the moving planet would encounter it at a certain apparent angle (aberration), and the force experienced would come from a point a little in advance of the sun. The rate of description of areas would thus tend to increase; whereas in reality it is constant. Hence the force of gravity, if it travel at all, does so with a speed far greater than that of light. It appears to be practically instantaneous. (Cf. "Modern Views of Electricity," § 126, end of chap. xii.) Again, anything like a retarding effect of the medium through which the planets move would constitute a tangential force, entirely un-directed towards the sun. Hence no such frictional or retarding force can appreciably exist. It is, however, conceivable that both these effects might occur and just neutralize each other. The neutralization is unlikely to be exact for all the planets; and the fact is, that no trace of either effect has as yet been discovered. (See also p. 176.)

The planets are, however, subject to forces not directed towards the sun, viz. their attractions for each other; and these perturbing forces do produce a slight discrepancy from Kepler's second law, but a discrepancy which is completely subject to calculation.

No. 2. Kepler's first law proves that this central force diminishes in the same proportion as the square of the distance increases.

To prove the connection between the inverse-square law of distance, and the travelling in a conic section with the centre of force in one focus (the other focus being empty), is not so simple. It obviously involves some geometry, and must therefore be left to properly armed students. But it may be useful to state that the inverse-square law of distance, although the simplest possible law for force emanating from a point or sphere, is not to be regarded as self-evident or as needing no demonstration. The force of a magnetic pole on a magnetized steel scrap, for instance, varies as the inverse cube of the distance; and the curve described by such a particle would be quite different from a conic section—it would be a definite class of spiral (called Cotes's spiral). Again, on an iron filing the force of a single pole might vary more nearly as the inverse fifth power; and so on. Even when the thing concerned is radiant in straight lines, like light, the law of inverse squares is not universally true. Its truth assumes, first, that the source is a point or sphere; next, that there is no reflection or refraction of any kind; and lastly, that the medium is perfectly transparent. The law of inverse squares by no means holds from a prairie fire for instance, or from a lighthouse, or from a street lamp in a fog.

Mutual perturbations, especially the pull of Jupiter, prevent the path of a planet from being really and truly an ellipse, or indeed from being any simple re-entrant curve. Moreover, when a planet possesses a satellite, it is not the centre of the planet which ever attempts to describe the Keplerian ellipse, but it is the common centre of gravity of the two bodies. Thus, in the case of the earth and moon, the point which really does describe a close attempt at an ellipse is a point displaced about 3000 miles from the centre of the earth towards the moon, and is therefore only 1000 miles beneath the surface.

No. 3. Kepler's third law proves that all the planets are acted on by the same kind of force; of an intensity depending on the mass of the sun.

The third law of Kepler, although it requires geometry to state and establish it for elliptic motion (for which it holds just as well as it does for circular motion), is very easy to establish for circular motion, by any one who knows about centrifugal force. If m is the mass of a planet, v its velocity, r the radius of its orbit, and T the time of describing it; $2\pi r = vT$, and the centripetal force needed to hold it in its orbit is

$$\frac{mv^2}{r} \text{ or } \frac{4\pi^2 mr}{T^2}$$

Now the force of gravitative attraction between the planet and the sun is

$$\frac{VmS}{r^2},$$

where v is a fixed quantity called the gravitation-constant, to be determined if possible by experiment once for all. Now, expressing the fact that the force of gravitation *is* the force holding the planet in, we write,

$$\frac{4\pi^2 mr}{T^2} = \frac{VmS}{r^2},$$

whence, by the simplest algebra,

$$\frac{r^3 mr}{T^2} = \frac{VS}{4\pi^2}.$$

The mass of the planet has been cancelled out; the mass of the sun remains, multiplied by the gravitation-constant, and is seen to be proportional to the cube of the distance divided by the square of the periodic time: a ratio, which is therefore the same for all planets controlled by the sun. Hence, knowing r and T for any single planet, the value of VS is known.

No. 4. So by knowing the length of year and distance of any planet from the sun, the sun's mass can be calculated, in terms of that of the earth.

No. 5. For the satellites, the force acting depends on the mass of *their* central body, a planet. Hence the mass of any planet possessing a satellite becomes known.

The same argument holds for any other system controlled by a central body—for instance, for the satellites of Jupiter; only instead of *S* it will be natural to write *J*, as meaning the mass of Jupiter. Hence, knowing *r* and *T* for any one satellite of Jupiter, the value of *VJ* is known.

Apply the argument also to the case of moon and earth. Knowing the distance and time of revolution of our moon, the value of *VE* is at once determined; *E* being the mass of the earth. Hence, *S* and *J*, and in fact the mass of any central body possessing a visible satellite, are now known in terms of *E*, the mass of the earth (or, what is practically the same thing, in terms of *V*, the gravitation-constant). Observe that so far none of these quantities are known absolutely. Their relative values are known, and are tabulated at the end of the Notes above, but the finding of their absolute values is another matter, which we must defer.

But, it may be asked, if Kepler's third law only gives us the mass of a *central* body, how is the mass of a *satellite* to be known? Well, it is not easy; the mass of no satellite is known with much accuracy. Their mutual perturbations give us some data in the case of the satellites of Jupiter; but to our own moon this method is of course inapplicable. Our moon perturbs at first sight nothing, and accordingly its mass is not even yet known with exactness. The mass of comets, again, is quite unknown. All that we can be sure of is that they are smaller than a certain limit, else they would perturb the planets they pass near. Nothing of this sort has ever been detected. They are themselves perturbed plentifully, but they perturb nothing; hence we learn that their mass is small. The mass of a comet may, indeed, be a few million or even billion tons; but that is quite small in astronomy.

But now it may be asked, surely the moon perturbs the earth, swinging it round their common centre of gravity, and really describing its own orbit about this point instead of about the earth's centre? Yes, that is so; and a more precise consideration of Kepler's third law enables us to make a fair approximation to the position of this common centre of gravity, and thus practically to "weigh the moon," i.e. to compare its mass with that of the earth; for their masses will be inversely as their respective distances from the common centre of gravity or balancing point—on the simple steel-yard principle.

Hitherto we have not troubled ourselves about the precise point about which the revolution occurs, but Kepler's third law is not precisely accurate unless it is attended to. The bigger the revolving body the greater is the discrepancy: and we see in the table preceding Lecture III., on page 57, that Jupiter exhibits an error which, though very slight, is greater than that of any of the other planets, when the sun is considered the fixed centre.

Let the common centre of gravity of earth and moon be displaced a distance x from the centre of the earth, then the moon's distance from the real centre of revolution is not r, but r-x; and the equation of centrifugal force to gravitative-attraction is strictly

$$\underline{4\pi^2}(r-x) = \underline{VE}\ ,$$

$T^2 \qquad\qquad r^2$

instead of what is in the text above; and this gives a slightly modified "third law." From this equation, if we have any distinct method of determining VE (and the next section gives such a method), we can calculate x and thus roughly weigh the moon, since

$$\frac{r-x}{r} = \frac{E}{\mathrm{E}+\mathrm{M}},$$

but to get anything like a reasonable result the data must be very precise.

No. 6. The force constraining the moon in her orbit is the same gravity as gives terrestrial bodies their weight and regulates the motion of projectiles.

Here we come to the Newtonian verification already several times mentioned; but because of its importance I will repeat it in other words. The hypothesis to be verified is that the force acting on the moon is the same kind of force as acts on bodies we can handle and weigh, and which gives them their weight. Now the weight of a mass m is commonly written mg, where g is the intensity of terrestrial gravity, a thing easily measured; being, indeed, numerically equal to twice the distance a stone drops in the first second of free fall. [See table p. 205.] Hence, expressing that the weight of a body is due to gravity, and remembering that the centre of the earth's attraction is distant from us by one earth's radius (R), we can write

$$mg = \frac{Vm\mathrm{E}}{\mathrm{R}^2},$$

or

$$VE = gR^2 = 95{,}522 \text{ cubic miles-per-second per second.}$$

But we already know vE, in terms of the moon's motion, as $4\pi^2 r^3/T^2$ approximately, [more accurately, see preceding note, this quantity is $V(E + M)$]; hence we can easily see if the two determinations of this quantity agree.[20]

All these deductions are fundamental, and may be considered as the foundation of the *Principia*. It was these that flashed upon Newton during that moment of excitement when he learned the real size of the earth, and discovered his speculations to be true.

The next are elaborations and amplifications of the theory, such as in ordinary times are left for subsequent generations of theorists to discover and work out.

Newton did not work out these remoter consequences of his theory completely by any means: the astronomical and mathematical world has been working them out ever since; but he carried the theory a great way, and here it is that his marvellous power is most conspicuous.

It is his treatment of No. 7, the perturbations of the moon, that perhaps most especially has struck all future mathematicians with amazement. No. 7, No. 14, No. 15, these are the most inspired of the whole.

No. 7. The moon is attracted not only by the earth, but by the sun also; hence its orbit is perturbed, and Newton calculated out the chief of these perturbations.

Now running through the perturbations (p. 203) in order:—The first is in parenthesis, because it is mere excentricity. It is not a true perturbation at all, and more properly belongs to Kepler.

(*a*) The first true perturbation is what Ptolemy called "the evection," the principal part of which is a periodic change in the ellipticity or excentricity of the moon's orbit, owing to the pull of the sun. It is a complicated matter, and Newton only partially solved it. I shall not attempt to give an account of it.

(*b*) The next, "the variation," is a much simpler affair. It is caused by the fact that as the moon revolves round the earth it is half the time nearer to the sun than the earth is, and so gets pulled more than the average, while for the other fortnight it is further from the sun than the earth is, and so gets pulled less. For the week during which it is changing from a decreasing half to a new moon it is moving in the direction of the extra pull, and hence becomes new sooner than would have been expected. All next week it is moving against the same extra pull, and so arrives at quadrature (half moon) somewhat late. For the next fortnight it is in the region of too little pull, the earth gets pulled more than it does; the effect of this is to hurry it up for the third week, so that the full moon occurs a little early, and to retard it for the fourth week, so that the decreasing half moon like the increasing half occurs behind time again. Thus each syzygy (as new and full are technically called) is too early; each quadrature is too late; the maximum hurrying and slackening force being felt at the octants, or intermediate 45° points.

(*c*) The "annual equation" is a fluctuation introduced into the other perturbations by reason of the varying distance of the disturbing body, the sun, at different seasons of the year. Its magnitude plainly depends simply on the excentricity of the earth's orbit.

Both these perturbations, (*b*) and (*c*), Newton worked out completely.

(*d*) and (*e*) Next come the retrogression of the nodes and the variation of the inclination, which at the time were being observed at Greenwich by Flamsteed, from whom Newton frequently, but vainly, begged for data that he might complete their theory while he had his mind upon it. Fortunately, Halley succeeded Flamsteed as Astronomer-Royal [see list at end of notes above], and then Newton would have no difficulty in gaining such information as the national Observatory could give.

The "inclination" meant is the angle between the plane of the moon's orbit and that of the earth. The plane of the earth's orbit round the sun is called the ecliptic; the plane of the moon's orbit round the earth is inclined to it at a certain angle, which is slowly changing, though in a periodic manner. Imagine a curtain ring bisected by a sheet of paper, and tilted to a certain angle; it may be likened to the moon's orbit, cutting the plane of the ecliptic. The two points at which the plane is cut by the ring are called "nodes"; and these nodes are not stationary, but are slowly regressing, *i.e.* travelling in a direction opposite to that of the moon itself. Also the angle of tilt is varying slowly, oscillating up and down in the course of centuries.

(*f*) The two points in the moon's elliptic orbit where it comes nearest to or farthest from the earth, *i.e.* the points at the extremity of the long axis of the ellipse, are called separately perigee and apogee, or together "the apses." Now the pull of the sun causes the whole orbit to slowly revolve in its own plane, and consequently these apses "progress," so that the true path is not quite a closed curve, but a sort of spiral with elliptic loops.

But here comes in a striking circumstance. Newton states with reference to this perturbation that theory only accounts for 1½° per annum, whereas observation gives 3°, or just twice as much.

This is published in the *Principia* as a fact, without comment. It was for long regarded as a very curious thing, and many great mathematicians afterwards tried to find an error in the working. D'Alembert, Clairaut, and others attacked the problem, but were led to just the same result. It constituted the great outstanding difficulty in the way of accepting the theory of gravitation. It was suggested that perhaps the inverse square law was only a first approximation; that perhaps a more complete expression, such as

$$\frac{A}{r^2} + \frac{B}{r^4},$$

must be given for it; and so on.

Ultimately, Clairaut took into account a whole series of neglected terms, and it came out correct; thus verifying the theory.

But the strangest part of this tale is to come. For only a few years ago, Prof. Adams, of Cambridge (Neptune Adams, as he is called), was editing various old papers of Newton's, now in the possession of the Duke of Portland, and he found manuscripts bearing on this very point, and discovered that Newton had reworked out the calculations himself, had found the cause of the error, had taken into account the terms hitherto neglected, and so, fifty years before Clairaut, had completely, though not publicly, solved this long outstanding problem of the progression of the apses.

(*g*) and (*h*) Two other inequalities he calculated out and predicted, viz. variation in the motions of the apses and the nodes. Neither of these had then been observed, but they were afterwards detected and verified.

A good many other minor irregularities are now known—some thirty, I believe; and altogether the lunar theory, or problem of the moon's exact motion, is one of the most complicated and difficult in astronomy; the perturbations being so numerous and large, because of the enormous mass of the perturbing body.

The disturbances experienced by the planets are much smaller, because they are controlled by the sun and perturbed by each other. The moon is controlled only by the earth, and perturbed by the sun. Planetary perturbations can be treated as a series of disturbances with some satisfaction: not so those of the moon. And yet it is the only way at present known of dealing with the lunar theory.

To deal with it satisfactorily would demand the solution of such a problem as this:—Given three rigid spherical masses thrown into empty space with any initial motions whatever, and abandoned to gravity: to determine their subsequent motions. With two masses the problem is simple enough, being pretty well summed up in Kepler's laws; but with three masses, strange to say, it is so complicated as to be beyond the reach of even modern mathematics. It is a famous problem, known as that of "the three bodies," but it has not yet been solved. Even when it is solved it will be only a close approximation to the case of earth, moon, and sun, for these bodies are not spherical, and are not rigid. One may imagine how absurdly and hopelessly complicated a complete treatment of the motions of the entire solar system would be.

No. 8. Each planet is attracted not only by the sun but by the other planets, hence their orbits are slightly affected by each other.

The subject of planetary perturbation was only just begun by Newton. Gradually (by Laplace and others) the theory became highly developed; and, as everybody knows, in 1846 Neptune was discovered by means of it.

No. 9. He recognized the comets as members of the solar system, obedient to the same law of gravity and moving in very elongated ellipses; so their return could be predicted.

It was a long time before Newton recognized the comets as real members of the solar system, and subject to gravity like the rest. He at first thought they moved in straight lines. It was only in the second edition of the *Principia* that the theory of comets was introduced.

Halley observed a fine comet in 1682, and calculated its orbit on Newtonian principles. He also calculated when it ought to have been seen in past times; and he found the year 1607, when one was seen by Kepler; also the year 1531, when one was seen by Appian; again, he reckoned 1456, 1380, 1305. All these appearances were the same comet, in all probability, returning every seventy-five or seventy-six years. The period was easily allowed to be not exact, because of perturbing planets. He then predicted its return for 1758, or perhaps 1759, a date he could not himself hope to see. He lived to a great age, but he died sixteen years before this date.

As the time drew nigh, three-quarters of a century afterwards, astronomers were greatly interested in this first cometary prediction, and kept an eager look-out for "Halley's comet." Clairaut, a most eminent mathematician and student of Newton, proceeded to calculate out more exactly the perturbing influence of Jupiter, near which it had passed. After immense labour (for the difficulty of the calculation was extreme, and the mass of mere figures something portentous), he predicted its return on the 13th of April, 1759, but he considered that he might have made a possible error of a month. It returned on the 13th of March, 1759, and established beyond all doubt the rule of the Newtonian theory over comets.

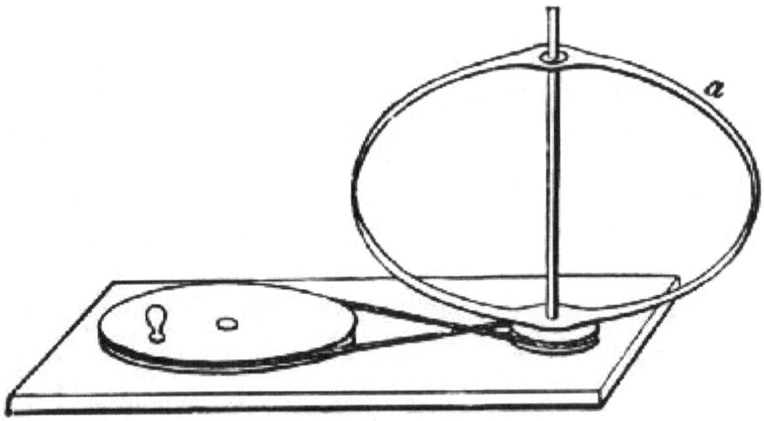

FIG. 71.—Well-known model exhibiting the oblate spheroidal form as a

consequence of spinning about a central axis. The brass strip *a* looks like a transparent globe when whirled, and bulges out equatorially.

No. 10. Applying the idea of centrifugal force to the earth considered as a rotating body, he perceived that it could not be a true sphere, and calculated its oblateness, obtaining 28 miles greater equatorial than polar diameter.

Here we return to one of the more simple deductions. A spinning body of any kind tends to swell at its circumference (or equator), and shrink along its axis (or poles). If the body is of yielding material, its shape must alter under the influence of centrifugal force; and if a globe of yielding substance subject to known forces rotates at a definite pace, its shape can be calculated. Thus a plastic sphere the size of the earth, held together by its own gravity, and rotating once a day, can be shown to have its equatorial diameter twenty-eight miles greater than its polar diameter: the two diameters being 8,000 and 8,028 respectively. Now we have no guarantee that the earth is of yielding material: for all Newton could tell it might be extremely rigid. As a matter of fact it is now very nearly rigid. But he argued thus. The water on it is certainly yielding, and although the solid earth might decline to bulge at the equator in deference to the diurnal rotation, that would not prevent the ocean from flowing from the poles to the equator and piling itself up as an equatorial ocean fourteen miles deep, leaving dry land everywhere near either pole. Nothing of this sort is observed: the distribution of land and water is not thus regulated. Hence, whatever the earth may be now, it must once have been plastic enough to accommodate itself perfectly to the centrifugal forces, and to take the shape appropriate to a perfectly plastic body. In all probability it was once molten, and for long afterwards pasty.

Thus, then, the shape of the earth can be calculated from the length of its day and the intensity of its gravity. The calculation is not difficult: it consists in imagining a couple of holes bored to the centre of the earth, one from a pole and one from the equator; filling these both with water, and calculating how much higher the water will stand in one leg of the gigantic V tube so formed than in the other. The answer comes out about fourteen miles.

The shape of the earth can now be observed geodetically, and it accords with calculation, but the observations are extremely delicate; in Newton's time the *size* was only barely known, the *shape* was not observed till long after; but on the principles of mechanics, combined with a little common-sense reasoning, it could be calculated with certainty and accuracy.

No. 11. From the observed shape of Jupiter or any planet the length of its day could be estimated.

Jupiter is much more oblate than the earth. Its two diameters are to one another as 17 is to 16; the ellipticity of its disk is manifest to simple inspection. Hence we perceive that its whirling action must be more violent—it must rotate quicker. As a matter of fact its day is ten

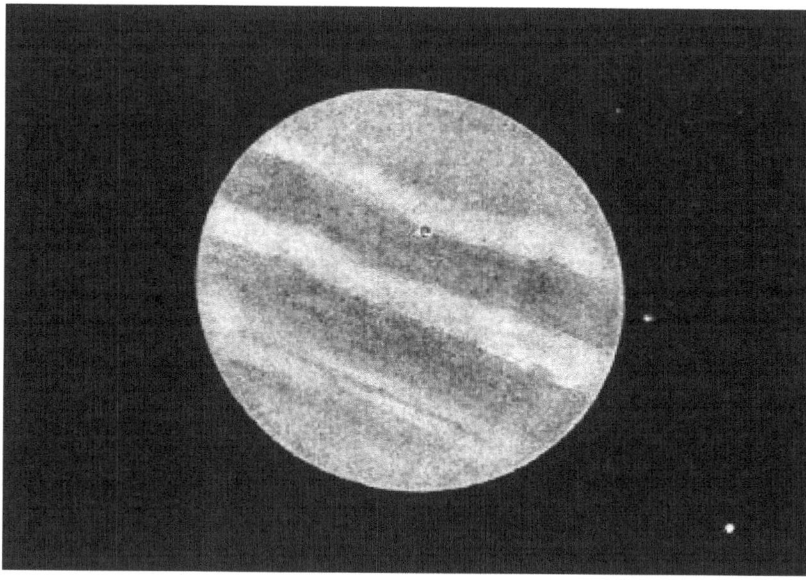

FIG. 72.—Jupiter.

hours long—five hours daylight and five hours night. The times of rotation of other bodies in the solar system are recorded in a table above.

No. 12. The so-calculated shape of the earth, in combination with centrifugal force, causes the weight of bodies to vary with latitude; and Newton calculated the amount of this variation. 194 lbs. at pole balance 195 lbs. at equator.

But following from the calculated shape of the earth follow several interesting consequences. First of all, the intensity of gravity will not be the same everywhere; for at the equator a stone is further from the average bulk of the earth (say the centre) than it is at the poles, and owing to this fact a mass of 590 pounds at the pole; would suffice to balance 591 pounds at the equator, if the two could be placed in the pans of a gigantic balance whose beam straddled along an earth's quadrant. This is a *true* variation of gravity due to the shape of the earth. But besides this there is a still larger *apparent* variation due to centrifugal force, which affects all bodies at the equator but not those at the poles. From this cause, even if the earth were a true sphere, yet if it were spinning at its actual pace, 288 pounds at the pole could balance 289 pounds at the equator; because at the equator the true weight of the mass would not be fully appreciated, centrifugal force would virtually diminish it by 1⁄289th of its amount.

In actual fact both causes co-exist, and accordingly the total variation of gravity observed is compounded of the real and the apparent effects; the result is that 194 pounds at a pole weighs as much as 195 pounds at the equator.

No. 13. A homogeneous sphere attracts as if its mass were concentrated at its centre. For any other figure, such as an oblate spheroid, this is not exactly true. A hollow concentric spherical shell exerts no force on small bodies inside it.

A sphere composed of uniform material, or of materials arranged in concentric strata, can be shown to attract external bodies as if its mass were concentrated at its centre. A hollow sphere, similarly composed, does the same, but on internal bodies it exerts no force at all.

Hence, at all distances above the surface of the earth, gravity decreases in inverse proportion as the square of the distance from the centre of the earth increases; but, if you descend a mine, gravity decreases in this case also as you leave the surface, though not at the same rate as when you went up. For as you penetrate the crust you get inside a concentric shell, which is thus powerless to act upon you, and the earth you are now outside is a smaller one. At what rate the force decreases depends on the distribution of density; if the density were uniform all through, the law of variation would be the direct distance, otherwise it would be more complicated. Anyhow, the intensity of gravity is a maximum at the surface of the earth, and decreases as you travel from the surface either up or down.

No. 14. The earth's equatorial protuberance, being acted on by the attraction of the sun and moon, must disturb its axis of rotation in a calculated manner; and thus is produced the precession of the equinoxes.

Here we come to a truly awful piece of reasoning. A sphere attracts as if its mass were concentrated at its centre (No. 12), but a spheroid does not. The earth is a spheroid, and hence it pulls and is pulled by the moon with a slightly uncentric attraction. In other words, the line of pull does not pass through its precise centre. Now when we have a spinning body, say a top, overloaded on one side so that gravity acts on it unsymmetrically, what happens? The axis of rotation begins to rotate cone-wise, at a pace which depends on the rate of spin, and on the shape and mass of the top, as well as on the amount and leverage of the overloading.

Newton calculated out the rapidity of this conical motion of the axis of the earth, produced by the slightly unsymmetrical pull of the moon, and found that it would complete a revolution in 26,000 years—precisely what was wanted to explain the precession of the equinoxes. In fact he had discovered the physical cause of that precession.

Observe that there were three stages in this discovery of precession:—

First, the observation by Hipparchus, that the nodes, or intersections of the earth's orbit (the sun's apparent orbit) with the plane of the equator, were not stationary, but slowly moved.

Second, the description of this motion by Copernicus, by the statement that it was due to a conical motion of the earth's axis of rotation about its centre as a fixed point.

Third, the explanation of this motion by Newton as due to the pull of the moon on the equatorial protuberance of the earth.

The explanation *could* not have been previously suspected, for the shape of the earth, on which the whole theory depends, was entirely unknown till Newton calculated it.

Another and smaller motion of a somewhat similar kind has been worked out since: it is due to the unsymmetrical attraction of the other planets for this same equatorial protuberance. It shows itself as a periodic change in the obliquity of the ecliptic, or so-called recession of the apses, rather than as a motion of the nodes.[21]

No. 15. The waters of the ocean are attracted towards the sun and moon on one side, and whirled a little farther away than the solid earth on the other side: hence Newton explained all the main phenomena of the tides.

And now comes another tremendous generalization. The tides had long been an utter mystery. Kepler likens the earth to an animal, and the tides to his breathings and inbreathings, and says they follow the moon.

Galileo chaffs him for this, and says that it is mere superstition to connect the moon with the tides.

Descartes said the moon pressed down upon the waters by the centrifugal force of its vortex, and so produced a low tide under it.

Everything was fog and darkness on the subject. The legend goes that an astronomer threw himself into the sea in despair of ever being able to explain the flux and reflux of its waters.

Newton now with consummate skill applied his theory to the effect of the moon upon the ocean, and all the main details of tidal action gradually revealed themselves to him.

He treated the water, rotating with the earth once a day, somewhat as if it were a satellite acted on by perturbing forces. The moon as it revolves round the earth is perturbed by the sun. The ocean as it revolves round the earth (being held on by gravitation just as the moon is) is perturbed by both sun and moon.

The perturbing effect of a body varies directly as its mass, and inversely as the cube of its distance. (The simple law of inverse square does not apply, because a perturbation is a differential effect: the satellite or ocean when nearer to the perturbing body than the rest of the earth, is attracted more, and when further off it is attracted less than is the main body of the earth; and it is these differences alone which constitute the perturbation.) The moon is the more powerful of the two perturbing bodies, hence the main tides are due to the moon; and its chief action is to cause a pair of low waves or oceanic humps, of gigantic area, to travel round the earth once in a lunar day, *i.e.* in about 24 hours and 50 minutes. The sun makes a similar but still lower pair of low elevations to travel round once in a solar day of 24 hours. And the combination of the two pairs of humps, thus periodically overtaking each other, accounts for the well-known spring and neap tides,—spring tides when their maxima agree, neap tides when the maximum of one coincides with the minimum of the other: each of which events happens regularly once a fortnight.

These are the main effects, but besides these there are the effects of varying distances and obliquity to be taken into account; and so we have a whole series of minor disturbances, very like those discussed in No. 7, under the lunar theory, but more complex still, because there are two perturbing bodies instead of only one.

The subject of the tides is, therefore, very recondite; and though one may give some elementary account of its main features, it will be best to defer this to a separate lecture (Lecture XVII).

I had better, however, here say that Newton did not limit himself to the consideration of the primary oceanic humps: he pursued the subject into geographical detail. He pointed out that, although the rise and fall of the tide at mid-ocean islands would be but small, yet on stretches of coast the wave would fling itself, and by its momentum would propel the waters, to a much greater height—for instance, 20 or 30 feet; especially in some funnel-shaped openings like the Bristol Channel and the Bay of Fundy, where the concentrated impetus of the water is enormous.

He also showed how the tidal waves reached different stations in successive regular order each day; and how some places might be fed with tide by two distinct channels; and that if the time of these channels happened to differ by six hours, a high tide might be arriving by one channel and a low tide by the other, so that the place would only feel the difference, and so have a very small observed rise and fall; instancing a port in China (in the Gulf of Tonquin) where that approximately occurs.

In fact, although his theory was not, as we now know, complete or final, yet it satisfactorily explained a mass of intricate detail as well as the main features of the tides.

No. 16. The sun's mass being known, he calculated the height of the solar tide.

No. 17. From the observed heights of spring and neap tides he determined the lunar tide, and thence made an estimate of the mass of the moon.

Knowing the sun's mass and distance, it was not difficult for Newton to calculate the height of the protuberance caused by it in a pasty ocean covering the whole earth. I say pasty, because, if there was any tendency for impulses to accumulate, as timely pushes given to a pendulum accumulate, the amount of disturbance might become excessive, and its calculation would involve a multitude of data. The Newtonian tide ignored this, thus practically treating the motion as either dead-beat, or else the impulses as very inadequately timed. With this reservation the mid-ocean tide due to the action of the sun alone comes out about one foot, or let us say one foot for simplicity. Now the actual tide observed in mid-Atlantic is at the springs about four feet, at the neaps about two. The spring tide is lunar plus solar; the neap tide is lunar minus solar. Hence it appears that the tide caused by the moon alone must be about three feet, when unaffected by momentum. From this datum Newton made the first attempt to approximately estimate the mass of the moon. I said that the masses of satellites must be estimated, if at all, by the perturbation they are able to cause. The lunar tide is a perturbation in the diurnal motion of the sea, and its amount is therefore a legitimate mode of calculating the moon's mass. The available data were not at all good, however; nor are they even now very perfect; and so the estimate was a good way out. It is now considered that the mass of the moon is about one-eightieth that of the earth.

Such are some of the gems extracted from their setting in the *Principia*, and presented as clearly as I am able before you.

Do you realize the tremendous stride in knowledge—not a stride, as Whewell says, nor yet a leap, but a flight—which has occurred between the dim gropings of Kepler, the elementary truths of Galileo, the fascinating but wild speculations of Descartes, and this magnificent and comprehensive system of ordered knowledge. To some his genius seemed almost divine. "Does Mr. Newton eat, drink, sleep, like other men?" said the Marquis de l'Hôpital, a French mathematician of no mean eminence; "I picture him to myself as a celestial genius, entirely removed from the restrictions of ordinary matter." To many it seemed as if there was nothing more to be discovered, as if the universe were now explored, and only a few fragments of truth remained for the gleaner. This is the attitude of mind expressed in Pope's famous epigram:—

> "Nature and Nature's laws lay hid in Night,
> God said, Let Newton be, and all was light."

This feeling of hopelessness and impotence was very natural after the advent of so overpowering a genius, and it prevailed in England for fully a century. It was very natural, but it was very mischievous; for, as a consequence, nothing of great moment was done by England in science, and no Englishman of the first magnitude appeared, till some who are either living now or who have lived within the present century.

It appeared to his contemporaries as if he had almost exhausted the possibility of discovery; but did it so appear to Newton? Did it seem to him as if he had seen far and deep into the truths of this great and infinite universe? It did not. When quite an old man, full of honour and renown, venerated, almost worshipped, by his contemporaries, these were his words:—

"I know not what the world will think of my labours, but to myself it seems that I have been but as a child playing on the sea-shore; now finding some pebble rather more polished, and now some shell rather more agreeably variegated than another, while the immense ocean of truth extended itself unexplored before me."

And so it must ever seem to the wisest and greatest of men when brought into contact with the great things of God—that which they know is as nothing, and less than nothing, to the infinitude of which they are ignorant.

Newton's words sound like a simple and pleasing echo of the words of that great unknown poet, the writer of the book of Job:—

> "Lo, these are parts of His ways,
> But how little a portion is heard of Him;
> The thunder of His power, who can
> understand?"

END OF PART I.

PART II
A COUPLE OF CENTURIES' PROGRESS.

NOTES TO LECTURE X

Science during the century after Newton

The *Principia* published, 1687

Roemer	1644–1710
James Bradley	1692–1762
Clairaut	1713–1765
Euler	1707–1783
D'Alembert	1717–1783
Lagrange	1736–1813
Laplace	1749–1827
William Herschel	1738–1822

Olaus Roemer was born in Jutland, and studied at Copenhagen. Assisted Picard in 1671 to determine the exact position of Tycho's observatory on Huen. Accompanied Picard to Paris, and in 1675 read before the Academy his paper "On Successive Propagation of Light as revealed by a certain inequality in the motion of Jupiter's First Satellite." In 1681 he returned to Copenhagen as Professor of Mathematics and Astronomy, and died in 1710. He invented the transit instrument, mural circle, equatorial mounting for telescopes, and most of the other principal instruments now in use in observatories. He made as many observations as Tycho Brahé, but the records of all but the work of three days were destroyed by a great fire in 1728.

Bradley, Professor of Astronomy at Oxford, discovered the aberration of light in 1729, while examining stars for parallax, and the nutation of the earth's axis in 1748. Was appointed Astronomer-Royal in 1742.

LECTURE X

ROEMER AND BRADLEY AND THE VELOCITY OF LIGHT

AT Newton's death England stood pre-eminent among the nations of Europe in the sphere of science. But the pre-eminence did not last long. Two great discoveries were made very soon after his decease, both by Professor Bradley, of Oxford, and then there came a gap. A moderately great man often leaves behind him a school of disciples able to work according to their master's methods, and with a healthy spirit of rivalry which stimulates and encourages them. Newton left, indeed, a school of disciples, but his methods of work were largely unknown to them, and such as were known were too ponderous to be used by ordinary men. Only one fresh result, and that a small one, has ever been attained by other men working according to the methods of the *Principia*. The methods were studied and commented on in England to the exclusion of all others for nigh a century, and as a consequence no really important work was done.

On the Continent, however, no such system of slavish imitation prevailed. Those methods of Newton's which had been simultaneously discovered by Leibnitz were more thoroughly grasped, modified, extended, and improved. There arose a great school of French and German mathematicians, and the laurels of scientific discovery passed to France and Germany—more especially, perhaps, at this time to France. England has never wholly recovered them. During the present century this country has been favoured with some giants who, as they become distant enough for their true magnitude to be perceived, may possibly stand out as great as any who have ever lived; but for the mass and bulk of scientific work at the present day we have to look to Germany, with its enlightened Government and extensive intellectual development. England, however, is waking up, and what its Government does not do, private enterprise is beginning to accomplish. The establishment of centres of scientific and literary activity in the great towns of England, though at present they are partially encumbered with the supply of education of an exceedingly rudimentary type, is a movement that in the course of another century or so will be seen to be one of the most important and fruitful steps ever taken by this country. On the Continent such centres have long existed; almost every large town is the seat of a University, and they are now liberally endowed. The University of Bologna (where, you may remember, Copernicus learnt mathematics) has recently celebrated its 800th anniversary.

The scientific history of the century after Newton, summarized in the above table of dates, embraces the labours of the great mathematicians Clairaut, Euler, D'Alembert, and especially of Lagrange and Laplace.

But the main work of all these men was hardly pioneering work. It was rather the surveying, and mapping out, and bringing into cultivation, of lands already discovered. Probably Herschel may be justly regarded as the next true pioneer. We shall not, however, properly appreciate the stages through which astronomy has passed, nor shall we be prepared adequately to welcome the discoveries of modern times unless we pay some attention to the intervening age. Moreover, during this era several facts of great moment gradually came into recognition; and the importance of the discovery we have now to speak of can hardly be over-estimated.

Our whole direct knowledge of the planetary and stellar universe, from the early observations of the ancients down to the magnificent discoveries of a Herschel, depends entirely upon our happening to possess a sense of sight. To no other of our senses do any other worlds than our own in the slightest degree appeal. We touch them or hear them never. Consequently, if the human race had happened to be blind, no other world but the one it groped its way upon could ever have been known or imagined by it. The outside universe would have existed, but man would have been entirely and hopelessly ignorant of it. The bare idea of an outside universe beyond the world would have been inconceivable, and might have been scouted as absurd. We do possess the sense of sight; but is it to be supposed that we possess every sense that can be possessed by finite beings? There is not the least ground for such an assumption. It is easy to imagine a deaf race or a blind race: it is not so easy to imagine a race more highly endowed with senses than our own; and yet the sense of smell in animals may give us some aid in thinking of powers of perception which transcend our own in particular directions. If there were a race with higher or other senses than our own, or if the human race should ever in the process of development acquire such extra sense-organs, a whole universe of existent fact might become for the first time perceived by us, and we should look back upon our past state as upon a blind chrysalid form of existence in which we had been unconscious of all this new wealth of perception.

It cannot be too clearly and strongly insisted on and brought home to every mind, that the mode in which the universe strikes us, our view of the universe, our whole idea of matter, and force, and other worlds, and even of consciousness, depends upon the particular set of sense-organs with which we, as men, happen to be endowed. The senses of force, of motion, of sound, of light, of touch, of heat, of taste, and of smell—these we have, and these are the things we primarily know. All else is inference founded upon these sensations. So the world appears to us. But given other sense-organs, and it might appear quite otherwise. What it is actually and truly like, therefore, is quite and for ever beyond us—so long as we are finite beings.

Without eyes, astronomy would be non-existent. Light it is which conveys all the information we possess, or, as it would seem, ever can possess, concerning the outer and greater universe in which this small world forms a speck. Light is the channel, the messenger of information; our eyes, aided by telescopes, spectroscopes, and many other "scopes" that may yet be invented, are the means by which we read the information that light brings.

Light travels from the stars to our eyes: does it come instantaneously? or does it loiter by the way? for if it lingers it is not bringing us information properly up to date—it is only telling us what the state of affairs was when it started on its long journey.

Now, it is evidently a matter of interest to us whether we see the sun as he is now, or only as he was some three hundred years ago. If the information came by express train it would be three hundred years behind date, and the sun might have gone out in the reign of

Queen Anne without our being as yet any the wiser. The question, therefore, "At what rate does our messenger travel?" is evidently one of great interest for astronomers, and many have been the attempts made to solve it. Very likely the ancient Greeks pondered over this question, but the earliest writer known to me who seriously discussed the question is Galileo. He suggests a rough experimental means of attacking it. First of all, it plainly comes quicker than sound. This can be perceived by merely watching distant hammering, or by noticing that the flash of a pistol is seen before its report is heard, or by listening to the noise of a flash of lightning. Sound takes five seconds to travel a mile—it has about the same speed as a rifle bullet; but light is much quicker than that.

The rude experiment suggested by Galileo was to send two men with lanterns and screens to two distant watch-towers or neighbouring mountain tops, and to arrange that each was to watch alternate displays and obscurations of the light made by the other, and to imitate them as promptly as possible. Either man, therefore, on obscuring or showing his own light would see the distant glimmer do the same, and would be able to judge if there was any appreciable interval between his own action and the response of the distant light. The experiment was actually tried by the Florentine Academicians,[22] with the result that, as practice improved, the interval became shorter and shorter, so that there was no reason to suppose that there was any real interval at all. Light, in fact, seemed to travel instantaneously.

Well might they have arrived at this result. Even if they had made far more perfect arrangements—for instance, by arranging a looking-glass at one of the stations in which a distant observer might see the reflection of his own lantern, and watch the obscurations and flashings made by himself, without having to depend on the response of human mechanism—even then no interval whatever could have been detected.

If, by some impossibly perfect optical arrangement, a lighthouse here were made visible to us after reflection in a mirror erected at New York, so that the light would have to travel across the Atlantic and back before it could be seen, even then the appearance of the light on removing a shutter, or the eclipse on interposing it, would seem to happen quite instantaneously. There would certainly be an interval: the interval would be the fiftieth part of a second (the time a stone takes to drop 1⁄13th of an inch), but that is too short to be securely detected without mechanism. With mechanism the thing might be managed, for a series of shutters might be arranged like the teeth of a large wheel; so that, when the wheel rotates, eclipses follow one another very rapidly; if then an eye looked through the same opening as that by which the light goes on its way to the distant mirror, a tooth might have moved sufficiently to cover up this space by the time the light returned; in which case the whole would appear dark, for the light would be stopped by a tooth, either at starting or at returning, continually. At higher speeds of rotation some light would reappear, and at lower speeds it would also reappear; by noticing, therefore, the precise speed at which there was constant eclipse the velocity of light could be determined.

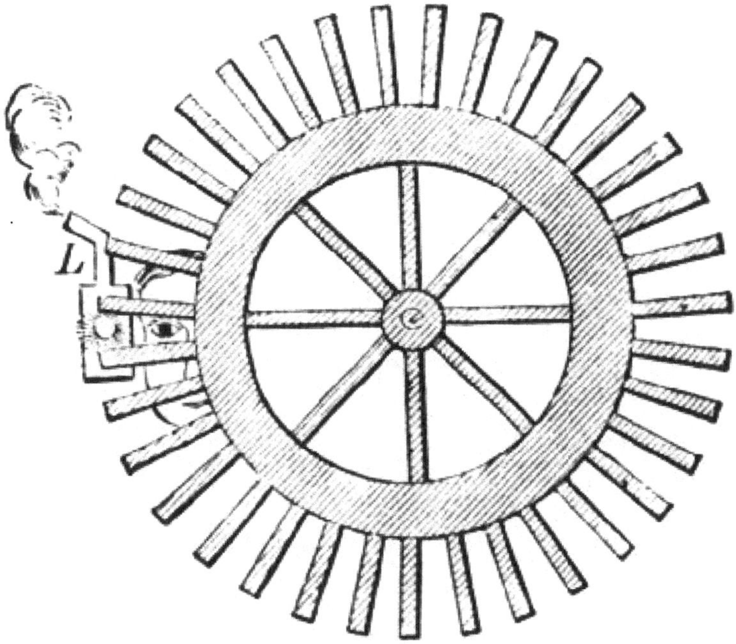

FIG. 73.—Diagram of eye looking at a light reflected in a distant mirror through the teeth of a revolving wheel.

This experiment has now been made in a highly refined form by Fizeau, and repeated by M. Cornu with prodigious care and accuracy. But with these recent matters we have no concern at present. It may be instructive to say, however, that if the light had to travel two miles altogether, the wheel would have to possess 450 teeth and to spin 100 times a second (at the risk of flying to pieces) in order that the ray starting through any one of the gaps might be stopped on returning by the adjacent tooth.

Well might the velocity of light be called instantaneous by the early observers. An ordinary experiment seemed (and was) hopeless, and light was supposed to travel at an infinite speed. But a phenomenon was noticed in the heavens by a quick-witted and ingenious Danish astronomer, which was not susceptible of any ordinary explanation, and which he perceived could at once be explained if light had a certain rate of travel—great, indeed, but something short of infinite. This phenomenon was connected with the satellites of Jupiter, and the astronomer's name was Roemer. I will speak first of the observation and then of the man.

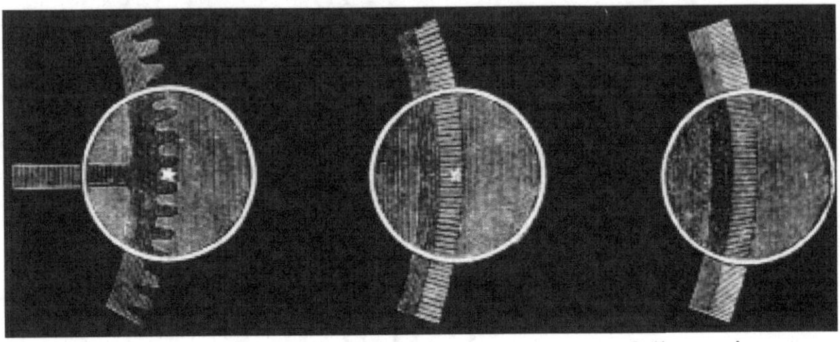

FIG. 74.—Fizeau's wheel, shewing the appearance of distant image seen through its teeth. 1st, when stationary, next when revolving at a moderate speed, last when revolving at the high speed just sufficient to cause eclipse.

Jupiter's satellites are visible, precisely as our own moon is, by reason of the shimmer of sunlight which they reflect. But as they revolve round their great planet they plunge into his shadow at one part of their course, and so become eclipsed from sunshine and invisible to us. The moment of disappearance can be sharply observed.

Take the first satellite as an example. The interval between successive eclipses ought to be its period of revolution round Jupiter. Observe this period. It was not uniform. On the average it was 42 hours 47 minutes, but it seemed to depend on the time of year. When Roemer observed in spring it was less, and in autumn it was more than usual. This was evidently a puzzling fact: what on earth can our year have to do with the motion of a moon of Jupiter's? It was probably, therefore, only an apparent change, caused either by our greater or less distance from Jupiter, or else by our greater or less speed of travelling to or from him. Considering it thus, he was led to see that, when the time of revolution seemed longest, we were receding fastest from Jupiter, and when shortest, approaching fastest.

If, then, light took time on its journey, *if* it travelled progressively, the whole anomaly would be explained.

In a second the earth goes nineteen miles; therefore in 42¾ hours (the time of revolution of Jupiter's first satellite) it goes 2·9 million (say three million) miles. The eclipse happens punctually, but we do not see it till the light conveying the information has travelled the extra three million miles and caught up the earth. Evidently, therefore, by observing how much the apparent time of revolution is lengthened in one part of the earth's orbit and shortened in another, getting all the data accurately, and assuming the truth of our hypothetical explanation, we can calculate the velocity of light. This is what Roemer did.

Now the maximum amount of retardation is just about fifteen seconds. Hence light takes this time to travel three million miles; therefore its velocity is three million divided by fifteen, say 200,000, or, as we now know more exactly, 186,000 miles every second. Note that the delay does not depend on our *distance*, but on our *speed*. One can tell this by common-sense as soon as we grasp the general idea of the explanation. A velocity cannot possibly depend on a distance only.

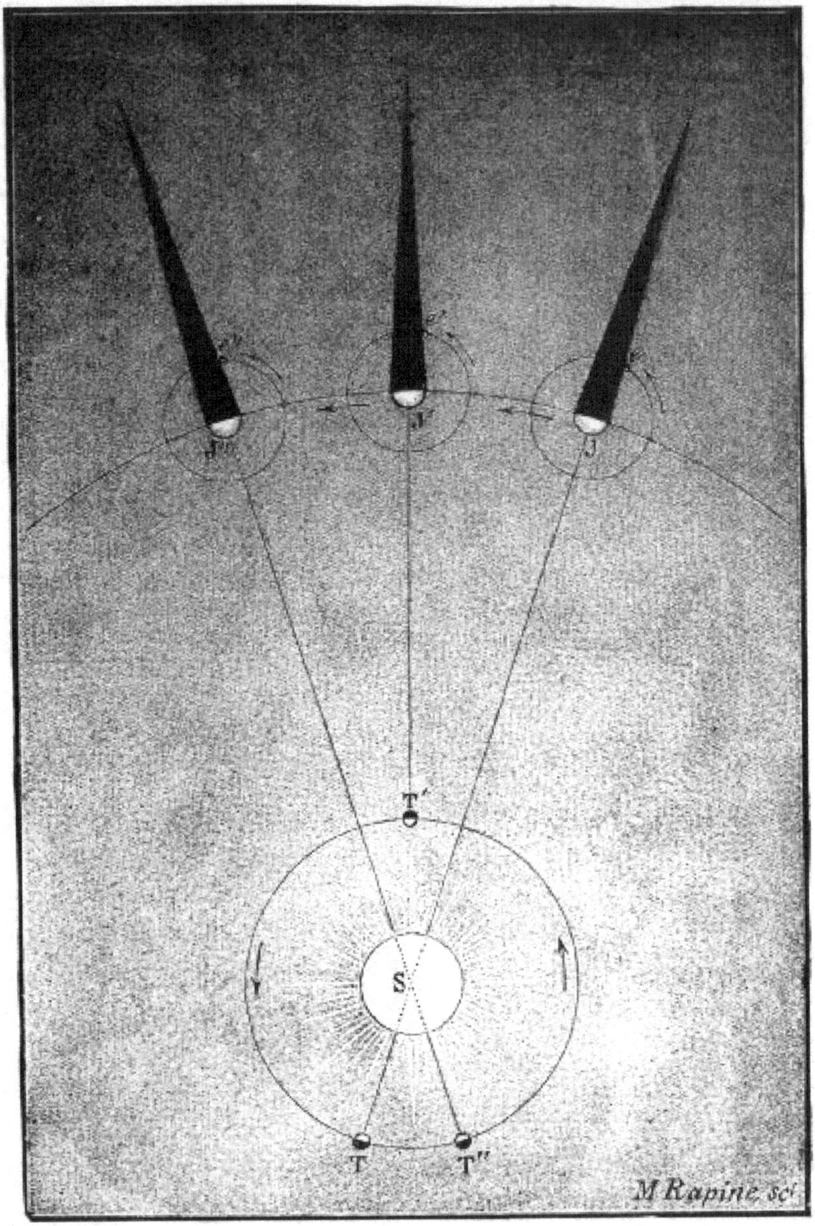

FIG. 75.—Eclipses of one of Jupiter's satellites. A diagram intended to illustrate the dependence of its apparent time of revolution (from

eclipse to eclipse) on the motion of the earth; but not illustrating the matter at all well. TT' T" are successive positions of the earth, while JJ' J" are corresponding positions of Jupiter.

Roemer's explanation of the anomaly was not accepted by astronomers. It excited some attention, and was discussed, but it was found not obviously applicable to any of the satellites except the first, and not very simply and satisfactorily even to that. I have, of course, given you the theory in its most elementary and simple form. In actual fact a host of disturbing and complicated considerations come in—not so violently disturbing for the first satellite as for the others, because it moves so quickly, but still complicated enough.

The fact is, the real motion of Jupiter's satellites is a most difficult problem. The motion even of our own moon (the lunar theory) is difficult enough: perturbed as its motion is by the sun. You know that Newton said it cost him more labour than all the rest of the *Principia*. But the motion of Jupiter's satellites is far worse. No one, in fact, has yet worked their theory completely out. They are perturbed by the sun, of course, but they also perturb each other, and Jupiter is far from spherical. The shape of Jupiter, and their mutual attractions, combine to make their motions most peculiar and distracting.

Hence an error in the time of revolution of a satellite was not *certainly* due to the cause Roemer suggested, unless one could be sure that the inequality was not a real one, unless it could be shown that the theory of gravitation was insufficient to account for it. This had not then been done; so the half-made discovery was shelved, and properly shelved, as a brilliant but unverified speculation. It remained on the shelf for half a century, and was no doubt almost forgotten.

FIG. 76.—A Transit-instrument for the British astronomical expedition, 1874. Shewing in its essential features the simplest form of such an instrument.

Now a word or two about the man. He was a Dane, educated at Copenhagen, and learned in the mathematics. We first hear of him as appointed to assist Picard, the eminent French geodetic surveyor (whose admirable work in determining the length of a degree you remember in connection with Newton), who had come over to Denmark with the object of fixing the exact site of the old and extinct Tychonic observatory in the island of Huen. For of course the knowledge of the exact latitude and longitude of every place whence numerous observations have been taken must be an essential to the full interpretation of those observations. The measurements being finished, young Roemer accompanied Picard to Paris, and here it was, a few years after, that he read his famous paper concerning "An Inequality in the Motion of Jupiter's First Satellite," and its explanation by means of an hypothesis of "the successive propagation of light."

The later years of his life he spent in Copenhagen as a professor in the University and an enthusiastic observer of the heavens,—not a descriptive observer like Herschel, but a measuring observer like Sir George Airy or Tycho Brahé. He was, in fact, a worthy follower of Tycho, and the main work of his life is the development and devising of new and more accurate astronomical instruments. Many of the large and accurate instruments with which a modern observatory is furnished are the invention of this Dane. One of the finest observatories in the world is the Russian one at Pulkowa, and a list of the instruments there reads like an extended catalogue of Roemer's inventions.

He not only *invented* the instruments, he had them made, being allowed money for the purpose; and he used them vigorously, so that at his death he left great piles of manuscript stored in the national observatory.

Unfortunately this observatory was in the heart of the city, and was thus exposed to a danger from which such places ought to be as far as possible exempt.

Some eighteen years after Roemer's death a great conflagration broke out in Copenhagen, and ruined large portions of the city. The successor to Roemer, Horrebow by name, fled from his house, with such valuables as he possessed, to the observatory, and there went on with his work. But before long the wind shifted, and to his horror he saw the flames coming his way. He packed up his own and his predecessor's manuscript observations in two cases, and prepared to escape with them, but the neighbours had resorted to the observatory as a place of safety, and so choked up the staircase with their property that he was barely able to escape himself, let alone the luggage, and everything was lost.

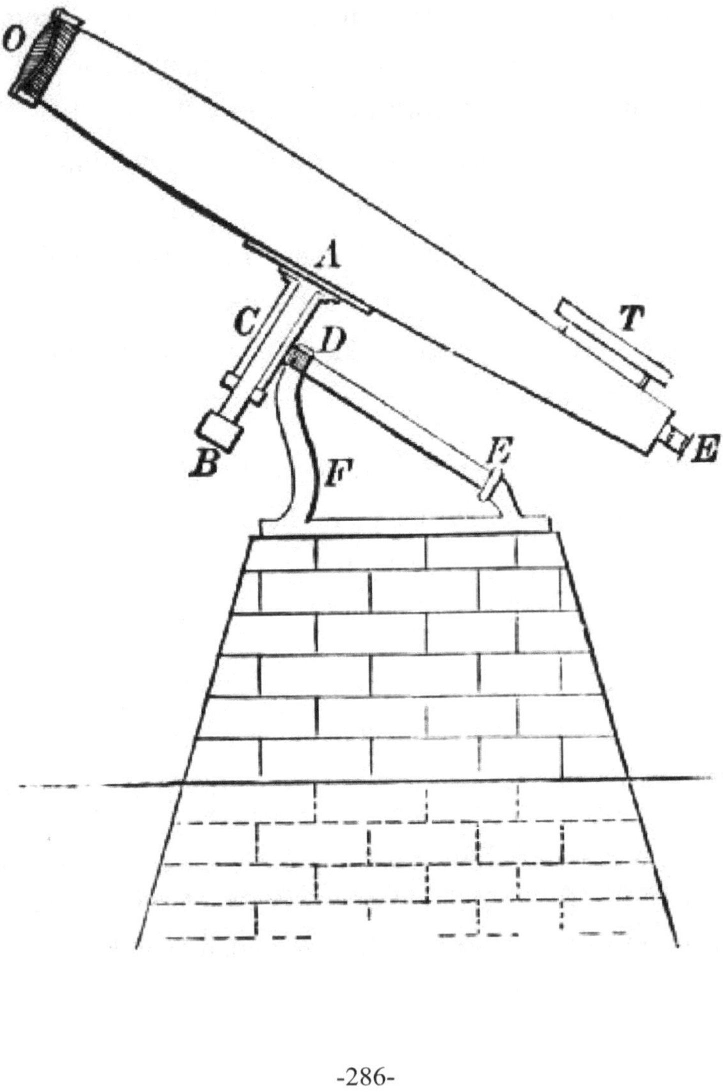

FIG. 77.—Diagram of equatorially mounted telescope; CE is the polar axis parallel to the axis of the earth; AB the declination axis. The diurnal motion is compensated by motion about the polar axis only, the other being clamped.

Of all the observations, only three days' work remains, and these were carefully discussed by Dr. Galle, of Berlin, in 1845, and their nutriment extracted. These ancient observations are of great use for purposes of comparison with the present state of the heavens, and throw light upon possible changes that are going on. Of course nowadays such a series of observations would be printed and distributed in many libraries, and so made practically indestructible.

Sad as the disaster was to the posthumous fame of the great observer, a considerable compensation was preparing. The very year that the fire occurred in Denmark a quiet philosopher in England was speculating and brooding on a remarkable observation that he had made concerning the apparent motion of certain stars, and he was led thereby to a discovery of the first magnitude concerning the speed of light—a discovery which resuscitated the old theory of Roemer about Jupiter's satellites, and made both it and him immortal.

James Bradley lived a quiet, uneventful, studious life, mainly at Oxford but afterwards at the National Observatory at Greenwich, of which he was third Astronomer-Royal, Flamsteed and Halley having preceded him in that office. He had taken orders, and lectured at Oxford as Savilian Professor. It is said that he pondered his great discovery while pacing the Long Walk at Magdalen College—and a beautiful place it is to meditate in.

Bradley was engaged in making observations to determine if possible the parallax of some of the fixed stars. Parallax means the apparent relative shift of bodies due to a change in the observer's position. It is parallax which we observe when travelling by rail and looking out of window at the distant landscape. Things at different distances are left behind at different apparent rates, and accordingly they seem to move relatively to each other. The most distant objects are least affected; and anything enormously distant, like the moon, is not subject to this effect, but would retain its position however far we travelled, unless we had some extraordinarily precise means of observation.

So with the fixed stars: they were being observed from a moving carriage—viz. the earth—and one moving at the rate of nineteen miles a second. Unless they were infinitely distant, or unless they were all at the same distance, they must show relative apparent motions among themselves. Seen from one point of the earth's orbit, and then in six months from an opposite point, nearly 184 million miles away, surely they must show some difference of aspect.

Remember that the old Copernican difficulty had never been removed. If the earth revolved round the sun, how came it that the fixed stars showed no parallax? The fact still remained a surprise, and the question a challenge. Picard, like other astronomers, supposed that it was only because the methods of observation had not been delicate enough; but now that, since the invention of the telescope and the founding of National Observatories, accuracy hitherto undreamt of was possible, why not attack the problem anew? This, then, he did, watching the stars with great care to see if in six months they showed any change in absolute position with reference to the pole of the heavens; any known secular motion of the pole, such as precession, being allowed for. Already he thought he detected a slight parallax for several stars near the pole, and the subject was exciting much interest.

Bradley determined to attempt the same investigation. He was not destined to succeed in it. Not till the present century was success in that most difficult observation achieved; and even now it cannot be done by the absolute methods then attempted; but, as so often happens, Bradley, in attempting one thing, hit upon another, and, as it happened, one of still greater brilliance and importance. Let us trace the stages of his discovery.

Atmospheric refraction made horizon observations useless for the delicacy of his purpose, so he chose stars near the zenith, particularly one—γ Draconis. This he observed very carefully at different seasons of the year by means of an instrument specially adapted for zenith observations, viz. a zenith sector. The observations were made in conjunction with a friend of his, an amateur astronomer named Molyneux, and they were made at Kew. Molyneux was shortly made First Lord of the Admiralty, or something important of that sort, and gave up frivolous pursuits. So Bradley observed alone. They observed the star accurately early in the month of December, and then intended to wait six months. But from curiosity Bradley observed it again only about a week later. To his surprise, he found that it had already changed its position. He recorded his observation on the back of an old envelope: it was his wont thus to use up odd scraps of paper—he was not, I regret to say, a tidy or methodical person—and this odd piece of paper turned up long afterwards among his manuscripts. It has been photographed and preserved as an historical relic.

Again and again he repeated the observation of the star, and continually found it moving still a little further and further south, an excessively small motion, but still an appreciable one—not to be set down to errors of observation. So it went on till March. It then waited, and after a bit longer began to return, until June. By September it was displaced as much to the north as it had been to the south, and by December it had got back to its original position. It had described, in fact, a small oscillation in the course of the year. The motion affected neighbouring stars in a similar way, and was called an "aberration," or wandering from their true place.

For a long time Bradley pondered over this observation, and over others like them which he also made. He found one group of stars describing small circles, while others at a distance from them were oscillating in straight lines, and all the others were describing ellipses. Unless this state of things were cleared up, accurate astronomy was impossible. The fixed stars!—they were not fixed a bit. To refined and accurate observation, such as was now possible, they were all careering about in little orbits having a reference to the earth's year, besides any proper motion which they might really have of their own, though no such motion was at present known. Not till Herschel was that discovered; not till this extraordinary aberration was allowed for could it be discovered. The effect observed by Bradley and Molyneux must manifestly be only an apparent motion: it was absurd to suppose a real stellar motion regulating itself according to the position of the earth. Parallax could not do it, for that would displace stars relatively among each other—it would not move similarly a set of neighbouring stars.

At length, four years after the observation, the explanation struck him, while in a boat upon the Thames. He noticed the apparent direction of the wind changed whenever the boat started. The wind veered when the boat's motion changed. Of course the cause of this was obvious enough—the speed of the wind and the speed of the boat were compounded, and gave an apparent direction of the wind other than the true direction. But this immediately suggested a cause for what he had observed in the heavens. He had been observing an apparent direction of the stars other than the true direction, because he was observing from a moving vehicle. The real direction was doubtless fixed: the apparent direction veered about with the motion of the earth. It must be that light did not travel instantaneously, but gradually, as Roemer had surmised fifty years ago; and that the motion of the light was compounded with the motion of the earth.

Think of a stream of light or anything else falling on a moving carriage. The carriage will run athwart the stream, the occupants of the carriage will mistake its true direction. A rifle fired through the windows of a railway carriage by a man at rest outside would make its perforations not in the true line of fire unless the train is stationary. If the train is moving, the line joining the holes will point to a place in advance of where the rifle is really located.

So it is with the two glasses of a telescope, the object-glass and eye-piece, which are pierced by the light; an astronomer, applying his eye to the tube and looking for the origin of the disturbance, sees it apparently, but not in its real position—its apparent direction is displaced in the direction of the telescope's motion; by an amount depending on the ratio of the velocity of the earth to the velocity of light, and on the angle between those two directions.

FIG. 78.—Aberration diagram. The light-ray L penetrates the object-glass of the moving telescope at O, but does not reach the eye-piece until the telescope has travelled to the second position. Consequently a moving telescope does not point out the true direction of the light, but aims at a point a little in advance.

But how minute is the displacement! The greatest effect is obtained when the two motions are at right angles to each other, *i.e.*

when the star seen is at right angles to the direction of the earth's motion, but even then it is only 20", or 1⁄180th part of a degree; one-ninetieth of the moon's apparent diameter. It could not be detected without a cross-wire in the telescope, and would only appear as a slight displacement from the centre of the field, supposing the telescope accurately pointed to the true direction.

But if this explanation be true, it at once gives a method of determining the velocity of light. The maximum angle of deviation, represented as a ratio of arc ÷ radius, amounts to

$$\frac{1}{180 \times 57\frac{1}{3}} - \cdot 0001 = \frac{1}{10{,}000}$$

(a gradient of 1 foot in two miles). In other words, the velocity of light must be 10,000 times as great as the velocity of the earth in its orbit. This amounts to a speed of 190,000 miles a second—not so very different from what Roemer had reckoned it in order to explain the anomalies of Jupiter's first satellite.

Stars in the direction in which the earth was moving would not be thus affected; there would be nothing in mere approach or recession to alter direction or to make itself in any way visible. Stars at right angles to the earth's line of motion would be most affected, and these would be all displaced by the full amount of 20 seconds of arc. Stars in intermediate directions would be displaced by intermediate amounts.

But the line of the earth's motion is approximately a circle round the sun, hence the direction of its advance is constantly though slowly changing, and in one year it goes through all the points of the compass. The stars, being displaced always in the line of advance, must similarly appear to describe little closed curves, always a quadrant in advance of the earth, completing their orbits once a year. Those near the pole of the ecliptic will describe circles, being always at right angles to the motion. Those in the plane of the ecliptic (near the zodiac) will be sometimes at right angles to the motion, but at other times will be approached or receded from; hence these will oscillate like pendulums once a year; and intermediate stars will have intermediate motions—that is to say, will describe ellipses of varying excentricity, but all completed in a year, and all with the major axis 20". This agreed very closely with what was observed.

The main details were thus clearly and simply explained by the hypothesis of a finite velocity for light, "the successive propagation of light in time." This time there was no room for hesitation, and astronomers hailed the discovery with enthusiasm.

Not yet, however, did Bradley rest. The finite velocity of light explained the major part of the irregularities he had observed, but not the whole. The more carefully he measured the amount of the deviation, the less completely accurate became its explanation.

There clearly was a small outstanding error or discrepancy; the stars were still subject to an unexplained displacement—not, indeed, a displacement that repeated itself every year, but one that went through a cycle of changes in a longer period.

The displacement was only about half that of aberration, and having a longer period was rather more difficult to detect securely. But the major difficulty was the fact that the two sorts of disturbances were co-existent, and the skill of disentangling them, and exhibiting the true and complete cause of each inequality, was very brilliant.

For nineteen years did Bradley observe this minor displacement, and in that time he saw it go through a complete cycle. Its cause was now clear to him; the nineteen-year period suggested the explanation. It is the period in which the moon goes through all her changes—a period known to the ancients as the lunar cycle, or Metonic cycle, and used by them to predict eclipses. It is still used for the first rough approximation to the prediction of eclipses, and to calculate Easter. The "Golden Number" of the Prayer-book is the number of the year in this cycle.

The cause of the second inequality, or apparent periodic motion of the stars, Bradley made out to be a nodding motion of the earth's axis.

The axis of the earth describes its precessional orbit or conical motion every 26,000 years, as had long been known; but superposed upon this great movement have now been detected minute nods, each with a period of nineteen years.

The cause of the nodding is completely accounted for by the theory of gravitation, just as the precession of the equinoxes was. Both disturbances result from the attraction of the moon on the non-spherical earth—on its protuberant equator.

"Nutation" is, in fact, a small perturbation of precession. The motion may be observed in a non-sleeping top. The slow conical motion of the top's slanting axis represents the course of precession.

Sometimes this path is loopy, and its little nods correspond to nutation.

The probable existence of some such perturbation had not escaped the sagacity of Newton, and he mentions something about it in the *Principia*, but thinks it too small to be detected by observation. He was thinking, however, of a solar disturbance rather than a lunar one, and this is certainly very small, though it, too, has now been observed.

Newton was dead before Bradley made these great discoveries, else he would have been greatly pleased to hear of them.

These discoveries of aberration and nutation, says Delambre, the great French historian of science, secure to their author a distinguished place after Hipparchus and Kepler among the astronomers of all ages and all countries.

NOTES TO LECTURE XI

Lagrange and *Laplace*, both tremendous mathematicians, worked very much in alliance, and completed Newton's work. The *Mécanique Céleste* contains the higher intricacies of astronomy mathematically worked out according to the theory of gravitation. They proved the solar system to be stable; all its inequalities being periodic, not cumulative. And Laplace suggested the "nebular hypothesis" concerning the origin of sun and planets: a hypothesis previously suggested, and to some extent, elaborated, by Kant.

A list of some of the principal astronomical researches of Lagrange and Laplace:—Libration of the moon. Long inequality of Jupiter and Saturn. Perturbations of Jupiter's satellites. Perturbations of comets. Acceleration of the moon's mean motion. Improved lunar theory. Improvements in the theory of the tides. Periodic changes in the form and obliquity of the earth's orbit. Stability of the solar system considered as an assemblage of rigid bodies subject to gravity.

The two equations which establish the stability of the solar system are:—

$$Sum\ (me^2\sqrt{d}) = constant,$$

and

$$Sum\ (m\ tan^2\theta\sqrt{d}) = constant;$$

where m is the mass of each planet, d its mean distance from the sun, e the excentricity of its orbit, and θ the inclination of its plane. However the expressions above formulated may change for individual planets, the sum of them for all the planets remains invariable.

The period of the variations in excentricity of the earth's orbit is 86,000 years; the period of conical revolution of the earth's axis is 25,800 years. About 18,000 years ago the excentricity was at a maximum.

LECTURE XI

LAGRANGE AND LAPLACE—THE STABILITY OF THE SOLAR SYSTEM, AND THE NEBULAR HYPOTHESIS

LAPLACE was the son of a small farmer or peasant of Normandy. His extraordinary ability was noticed by some wealthy neighbours, and by them he was sent to a good school. From that time his career was one brilliant success, until in the later years of his life his prominence brought him tangibly into contact with the deteriorating influence of politics. Perhaps one ought rather to say trying than deteriorating; for they seem trying to a strong character, deteriorating to a weak one—and unfortunately, Laplace must be classed in this latter category.

It has always been the custom in France for its high scientific men to be conspicuous also in politics. It seems to be now becoming the fashion in this country also, I regret to say.

The *life* of Laplace is not specially interesting, and I shall not go into it. His brilliant mathematical genius is unquestionable, and almost unrivalled. He is, in fact, generally considered to come in this respect next after Newton. His talents were of a more popular order than those of Lagrange, and accordingly he acquired fame and rank, and rose to the highest dignities. Nevertheless, as a man and a politician he hardly commands our respect, and in time-serving adjustability he is comparable to the redoubtable Vicar of Bray. His scientific insight and genius were however unquestionably of the very highest order, and his work has been invaluable to astronomy.

I will give a short sketch of some of his investigations, so far as they can be made intelligible without overmuch labour. He worked very much in conjunction with Lagrange, a more solid though a less brilliant man, and it is both impossible and unnecessary for us to attempt to apportion respective shares of credit between these two scientific giants, the greatest scientific men that France ever produced.

First comes a research into the libration of the moon. This was discovered by Galileo in his old age at Arcetri, just before his blindness. The moon, as every one knows, keeps the same face to the earth as it revolves round it. In other words, it does not rotate with reference to the earth, though it does rotate with respect to outside bodies. Its libration consists in a sort of oscillation, whereby it shows us now a little more on one side, now a little more on the other, so that altogether we are cognizant of more than one-half of its surface—in fact, altogether of about three-fifths. It is a simple and unimportant matter, easily explained.

The motion of the moon may be analyzed into a rotation about its own axis combined with a revolution about the earth. The speed of the rotation is quite uniform, the speed of the revolution is not quite uniform, because the orbit is not circular but elliptical, and the moon has to travel faster in perigee than in apogee (in accordance with Kepler's second law). The consequence of this is that we see a little too far round the body of the moon, first on one side, then on the other. Hence it *appears* to oscillate slightly, like a lop-sided fly-wheel whose revolutions have been allowed to die away so that they end in oscillations of small amplitude.[23] Its axis of rotation, too, is not precisely perpendicular to its plane of revolution, and therefore we sometimes see a few hundred miles beyond its north pole, sometimes a similar amount beyond its south. Lastly, there is a sort of parallax effect, owing to the fact that we see the rising moon from one point of view, and the setting moon from a point 8,000 miles distant; and this base-line of the earth's diameter gives us again some extra glimpses. This diurnal or parallactic libration is really more effective than the other two in extending our vision into the space-facing hemisphere of the moon.

These simple matters may as well be understood, but there is nothing in them to dwell upon. The far side of the moon is probably but little worth seeing. Its features are likely to be more blurred with accumulations of meteoric dust than are those of our side, but otherwise they are likely to be of the same general character.

The thing of real interest is the fact that the moon does turn the same face towards us; *i.e.* has ceased to rotate with respect to the earth (if ever it did so). The stability of this state of things was shown by Lagrange to depend on the shape of the moon. It must be slightly egg-shape, or prolate—extended in the direction of the earth; its earth-pointing diameter being a few hundred feet longer than its visible diameter; a cause slight enough, but nevertheless sufficient to maintain stability, except under the action of a distinct disturbing cause. The prolate or lemon-like shape is caused by the gravitative pull of the earth, balanced by the centrifugal whirl. The two forces balance each other as regards motion, but between them they have strained the moon a trifle out of shape. The moon has yielded as if it were perfectly plastic; in all probability it once was so.

It may be interesting to note for a moment the correlative effect of this aspect of the moon, if we transfer ourselves to its surface in imagination, and look at the earth (cf. Fig. 41). The earth would be like a gigantic moon of four times our moon's diameter, and would go through its phases in regular order. But it would not rise or set: it would be fixed in the sky, and subject only to a minute oscillation to and fro once a month, by reason of the "libration" we have been speaking of. Its aspect, as seen by markings on its surface, would rapidly change, going through a cycle in twenty-four hours; but its permanent features would be usually masked by lawless accumulations of cloud, mainly aggregated in rude belts parallel to the equator. And these cloudy patches would be the most luminous, the whitest portions; for of course it would be their silver lining that we would then be looking on.[24]

Next among the investigations of Lagrange and Laplace we will mention the long inequality of Jupiter and Saturn. Halley had found that Jupiter was continually lagging behind its true place as given by the theory of gravitation; and, on the other hand, that Saturn was being accelerated. The lag on the part of Jupiter amounted to about 34½ minutes in a century. Overhauling ancient observations, however, Halley found signs of the opposite state of things, for when he got far enough back Jupiter was accelerated and Saturn was being retarded.

Here was evidently a case of planetary perturbation, and Laplace and Lagrange undertook the working of it out. They attacked it as a case of the problem of three bodies, viz. the sun, Jupiter, and Saturn; which are so enormously the biggest of the known bodies in the system that insignificant masses like the Earth, Mars, and the rest, may be wholly neglected. They succeeded brilliantly, after a long and complex investigation: succeeded, not in solving the problem of the three bodies, but, by considering their mutual action as perturbations superposed on each other, in explaining the most conspicuous of the observed anomalies of their motion, and in laying the foundation of a general planetary theory.

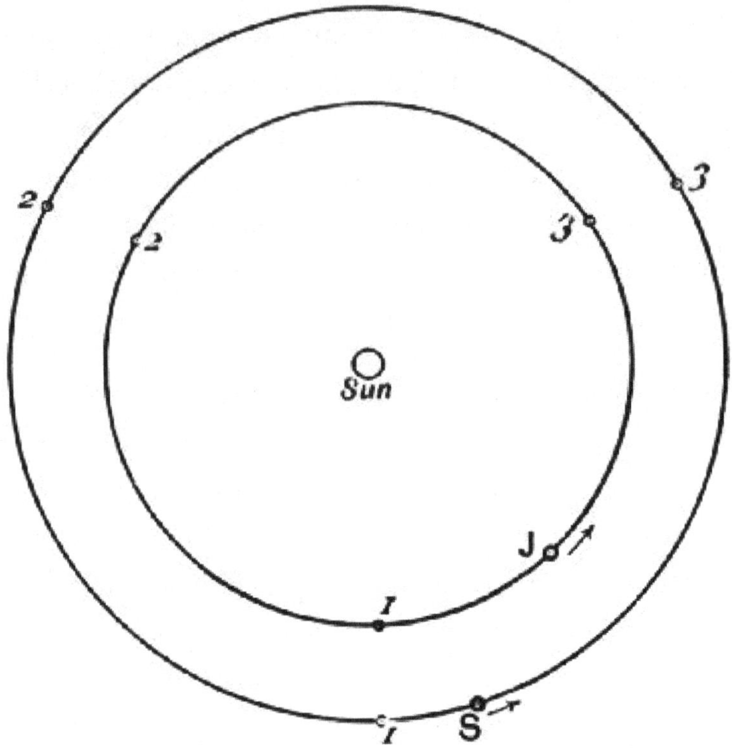

FIG. 79.—Shewing the three conjunction places in the orbits of Jupiter and Saturn. The two planets are represented as leaving one of the conjunctions where Jupiter was being pulled back and Saturn being pulled forward by their mutual attraction.

One of the facts that plays a large part in the result was known to the old astrologers, viz. that Jupiter and Saturn come into conjunction with a certain triangular symmetry; the whole scheme being called a trigon, and being mentioned several times by Kepler. It happens that five of Jupiter's years very nearly equal two of Saturn's,[25] so that they get very nearly into conjunction three times in every five Jupiter years, but not exactly. The result of this close approach is that periodically one pulls the other on and is itself pulled back; but since the three points progress, it is not always the same planet which gets pulled back. The complete theory shows that in the year 1560 there was no marked perturbation: before that it was in one direction, while afterwards it was in the other direction, and the period of the whole cycle of disturbances is 929 of our years. The solution of this long outstanding puzzle by the theory of gravitation was hailed with the greatest enthusiasm by astronomers, and it established the fame of the two French mathematicians.

Next they attacked the complicated problem of the motions of Jupiter's satellites. They succeeded in obtaining a theory of their motions which represented fact very nearly indeed, and they detected the following curious relationship between the satellites:—The speed of the first satellite + twice the speed of the second is equal to the speed of the third.

They found this, not empirically, after the manner of Kepler, but as a deduction from the law of gravitation; for they go on to show that even if the satellites had not started with this relation they would sooner or later, by mutual perturbation, get themselves into it. One singular consequence of this, and of another quite similar connection between their positions, is that all three satellites can never be eclipsed at once.

The motion of the fourth satellite is less tractable; it does not so readily form an easy system with the others.

After these great successes the two astronomers naturally proceeded to study the mutual perturbations of all other bodies in the solar system. And one very remarkable discovery they made concerning the earth and moon, an account of which will be interesting, though the details and processes of calculation are quite beyond us in a course like this.

Astronomical theory had become so nearly perfect by this time, and observations so accurate, that it was possible to calculate many astronomical events forwards or backwards, over even a thousand years or more, with admirable precision.

Now, Halley had studied some records of ancient eclipses, and had calculated back by means of the lunar theory to see whether the calculation of the time they ought to occur would agree with the record of the time they did occur. To his surprise he found a discrepancy, not a large one, but still one quite noticeable. To state it as we know it now:—An eclipse a century ago happened twelve seconds later than it ought to have happened by theory; two centuries back the error amounted to forty-eight seconds, in three centuries it would be 108 seconds, and so on; the lag depending on the square of the time. By research, and help from scholars, he succeeded in obtaining the records of some very ancient eclipses indeed. One in Egypt towards the end of the tenth century A.D.; another in 201 A.D.; another a little before Christ; and one, the oldest of all of which any authentic record has been preserved, observed by the Chaldæan astronomers in Babylon in the reign of Hezekiah.

Calculating back to this splendid old record of a solar eclipse, over the intervening 2,400 years, the calculated and the observed times were found to disagree by nearly two hours. Pondering over an explanation of the discrepancy, Halley guessed that it must be because the moon's motion was not uniform, it must be going quicker and quicker, gaining twelve seconds each century on its previous gain—a discovery announced by him as "the acceleration of the moon's mean motion." The month was constantly getting shorter.

What was the physical cause of this acceleration according to the theory of gravitation? Many attacked the question, but all failed. This was the problem Laplace set himself to work out. A singular and beautiful result rewarded his efforts.

You know that the earth describes an elliptic orbit round the sun: and that an ellipse is a circle with a certain amount of flattening or "excentricity."[26] Well, Laplace found that the excentricity of the earth's orbit must be changing, getting slightly less; and that this change of excentricity would have an effect upon the length of the month. It would make the moon go quicker.

One can almost see how it comes about. A decrease in excentricity means an increase in mean distance of the earth from the sun. This means to the moon a less solar perturbation. Now one effect of the solar perturbation is to keep the moon's orbit extra large: if the size of its orbit diminishes, its velocity must increase, according to Kepler's third law.

Laplace calculated the amount of acceleration so resulting, and found it ten seconds a century; very nearly what observation required; for, though I have quoted observation as demanding twelve seconds per century, the facts were not then so distinctly and definitely ascertained.

This calculation for a long time seemed thoroughly satisfactory, but it is not the last word on the subject. Quite lately an error has been found in the working, which diminishes the theoretical gravitation-acceleration to six seconds a century instead of ten, thus making it insufficient to agree exactly with fact. The theory of gravitation leaves an outstanding error. (The point is now almost thoroughly understood, and we shall return to it in Lecture XVIII).

But another question arises out of this discussion. I have spoken of the excentricity of the earth's orbit as decreasing. Was it always decreasing? and if so, how far back was it so excentric that at perihelion the earth passed quite near the sun? If it ever did thus pass near the sun, the inference is manifest—the earth must at one time have been thrown off, or been separated off, from the sun.

If a projectile could be fired so fast that it described an orbit round the earth—and the speed of fire to attain this lies between five and seven miles a second (not less than the one, nor more than the other)—it would ever afterwards pass through its point of projection as one point of its elliptic orbit; and its periodic return through that point would be the sign of its origin. Similarly, if a satellite does *not* come near its central orb, and can be shown never to have been near it, the natural inference is that it has *not* been born from it, but has originated in some other way.

The question which presented itself in connexion with the variable ellipticity of the earth's orbit was the following:—Had it always been decreasing, so that once it was excentric enough just to graze the sun at perihelion as a projected body would do?

Into the problem thus presented Lagrange threw himself, and he succeeded in showing that no such explanation of the origin of the earth is possible. The excentricity of the orbit, though now decreasing, was not always decreasing; ages ago it was increasing: it passes through periodic changes. Eighteen thousand years ago its excentricity was a maximum; since then it has been diminishing, and will continue to diminish for 25,000 years more, when it will be an almost perfect circle; it will then begin to increase again, and so on. The obliquity of the ecliptic is also changing periodically, but not greatly: the change is less than three degrees.

This research has, or ought to have, the most transcendent interest for geologists and geographers. You know that geologists find traces of extraordinary variations of temperature on the surface of the earth. England was at one time tropical, at another time glacial. Far away north, in Spitzbergen, evidence of the luxuriant vegetation of past ages has been found; and the explanation of these great climatic changes has long been a puzzle. Does not the secular variation in excentricity of the earth's orbit, combined with the precession of the equinoxes, afford a key? And if a key at all, it will be an accurate key, and enable us to calculate back with some precision to the date of the glacial epoch; and again to the time when a tropical flora flourished in what is now northern Europe, *i.e.* to the date of the Carboniferous era.

This aspect of the subject has recently been taught with vigour and success by Dr. Croll in his book "Climate and Time."

A brief and partial explanation of the matter may be given, because it is a point of some interest and is also one of fair simplicity.

Every one knows that the climatic conditions of winter and summer are inverted in the two hemispheres, and that at present the sun is nearest to us in our (northern) winter. In other words, the earth's axis is inclined so as to tilt its north pole away from the sun at perihelion, or when the earth is at the part of its elliptic orbit nearest the sun's focus; and to tilt it towards the sun at aphelion. The result of this present state of things is to diminish the intensity of the average northern winter and of the average northern summer, and on the other hand to aggravate the extremes of temperature in the southern hemisphere; all other things being equal. Of course other things are not equal, and the distribution of land and sea is a still more powerful climatic agent than is the three million miles or so extra nearness of the sun. But it is supposed that the Antarctic ice-cap is larger than the northern, and increased summer radiation with increased winter cold would account for this.

But the present state of things did not always obtain. The conical movement of the earth's axis (now known by a curious perversion of phrase as "precession") will in the course of 13,000 years or so cause the tilt to be precisely opposite, and then we shall have the more extreme winters and summers instead of the southern hemisphere.

If the change were to occur now, it might not be overpowering, because now the excentricity is moderate. But if it happened some time back, when the excentricity was much greater, a decidedly different arrangement of climate may have resulted. There is no need to say *if* it happened some time back: it did happen, and accordingly an agent for affecting the distribution of mean temperature on the earth is to hand; though whether it is sufficient to achieve all that has been observed by geologists is a matter of opinion.

Once more, the whole diversity of the seasons depends on the tilt of the earth's axis, the 23° by which it is inclined to a perpendicular to the orbital plane; and this obliquity or tilt is subject to slow fluctuations.

Hence there will come eras when all causes combine to produce a maximum extremity of seasons in the northern hemisphere, and other eras when it is the southern hemisphere which is subject to extremes.

But a grander problem still awaited solution—nothing less than the fate of the whole solar system. Here are a number of bodies of various sizes circulating at various rates round one central body, all attracted by it, and all attracting each other, the whole abandoned to the free play of the force of gravitation: what will be the end of it all? Will they ultimately approach and fall into the sun, or will they recede further and further from him, into the cold of space? There is a third possible alternative: may they not alternately approach and recede from him, so as on the whole to maintain a fair approximation to their present distances, without great and violent extremes of temperature either way?

If any one planet of the system were to fall into the sun, more especially if it were a big one like Jupiter or Saturn, the heat produced would be so terrific that life on this earth would be destroyed, even at its present distance; so that we are personally interested in the behaviour of the other planets as well as in the behaviour of our own.

The result of the portentously difficult and profoundly interesting investigation, here sketched in barest outline, is that the solar system is stable: that is to say, that if disturbed a little it will oscillate and return to its old state; whereas if it were unstable the slightest disturbance would tend to accumulate, and would sooner or later bring about a catastrophe. A hanging pendulum is stable, and oscillates about a mean position; its motion is periodic. A top-heavy load balanced on a point is unstable. All the changes of the solar system are periodic, *i.e.* they repeat themselves at regular intervals, and they never exceed a certain moderate amount.

The period is something enormous. They will not have gone through all their changes until a period of 2,000,000 years has elapsed. This is the period of the planetary oscillation: "a great pendulum of eternity which beats ages as our pendulums beat seconds." Enormous it seems; and yet we have reason to believe that the earth has existed through many such periods.

The two laws of stability discovered and stated by Lagrange and Laplace I can state, though they may be difficult to understand:—

Represent the masses of the several planets by m_1, m_2, &c.; their mean distances from the sun (or radii vectores) by r_1, r_2, &c.; the excentricities of their orbits by e_1, e_2, &c.; and the obliquity of the planes of these orbits, reckoned from a single plane of reference or "invariable plane," by θ_1, θ_2, &c.; then all these quantities (except m) are liable to fluctuate; but, however much they change, an increase for one planet will be accompanied by a decrease for some others; so that, taking all the planets into account, the sum of a set of terms like these, $m_1e_2^2\sqrt{r_1} + m_2e_2^2\sqrt{r_2} +$ &c., will remain always the same. This is summed up briefly in the following statement:

$$\Sigma(me^2\sqrt{r}) = \text{constant.}$$

That is one law, and the other is like it, but with inclination of orbit instead of excentricity, viz.:

$$\Sigma(m\theta^2\sqrt{r}) = \text{constant.}$$

The value of each of these two constants can at any time be calculated. At present their values are small. Hence they always were and always will be small; being, in fact, invariable. Hence neither e nor r nor θ can ever become infinite, nor can their average value for the system ever become zero.

The planets may share the given amount of total excentricity and obliquity in various proportions between themselves; but even if it were all piled on to one planet it would not be very excessive, unless the planet were so small a one as Mercury; and it would be most improbable that one planet should ever have all the excentricity of the solar system heaped upon itself. The earth, therefore, never has been, nor ever will be, enormously nearer the sun than it is at present: nor can it ever get very much further off. Its changes are small and are periodic—an increase is followed by a decrease, like the swing of a pendulum.

The above two laws have been called the Magna Charta of the solar system, and were long supposed to guarantee its absolute permanence. So far as the theory of gravitation carries us, they do guarantee its permanence; but something more remains to be said on the subject in a future lecture (XVIII).

And now, finally, we come to a sublime speculation, thrown out by Laplace, not as the result of profound calculation, like the results hitherto mentioned, not following certainly from the theory of gravitation, or from any other known theory, and therefore not to be accepted as more than a brilliant hypothesis, to be confirmed or rejected as our knowledge extends. This speculation is the "Nebular hypothesis." Since the time of Laplace the nebular hypothesis has had ups and downs of credence, sometimes being largely believed in, sometimes being almost ignored. At the present time it holds the field with perhaps greater probability of ultimate triumph than has ever before seemed to belong to it—far greater than belonged to it when first propounded.

It had been previously stated clearly and well by the philosopher Kant, who was intensely interested in "the starry heavens" as well as in the "mind of man," and who shewed in connexion with astronomy also a most surprising genius. The hypothesis ought by rights perhaps to be known rather by his name than by that of Laplace.

The data on which it was founded are these:—Every motion in the solar system known at that time took place in one direction, and in one direction only. Thus the planets revolve round the sun, all going the same way round; moons revolve round the planets, still maintaining the same direction of rotation, and all the bodies that were known to rotate on their own axis did so with still the same kind of spin. Moreover, all these motions take place in or near a single plane. The ancients knew that sun moon and planets all keep near to the ecliptic, within a belt known as the zodiac: none strays away into other parts of the sky. Satellites also, and rings, are arranged in or near the same plane; and the plane of diurnal spin, or equator of the different bodies, is but slightly tilted.

Now all this could not be the result of chance. What could have caused it? Is there any connection or common ancestry possible, to account for this strange family likeness? There is no connection now, but there may have been once. Must have been, we may almost say. It is as though they had once been parts of one great mass rotating as a whole; for if such a rotating mass broke up, its parts would retain its direction of rotation. But such a mass, filling all space as far as or beyond Saturn, although containing the materials of the whole solar system in itself, must have been of very rare consistency. Occupying so much bulk it could not have been solid, nor yet liquid, but it might have been gaseous.

Are there any such gigantic rotating masses of gas in the heaven now? Certainly there are; there are the nebulæ. Some of the nebulæ are now known to be gaseous, and some of them at least are in a state of rotation. Laplace could not have known this for certain, but he suspected it. The first distinctly spiral nebula was discovered by the telescope of Lord Rosse; and quite recently a splendid photograph of the great Andromeda nebula, by our townsman, Mr. Isaac Roberts, reveals what was quite unsuspected—and makes it clear that this prodigious mass also is in a state of extensive and majestic whirl.

Very well, then, put this problem:—A vast mass of rotating gas is left to itself to cool for ages and to condense as it cools: how will it behave? A difficult mathematical problem, worthy of being attacked to-day; not yet at all adequately treated. There are those who believe that by the complete treatment of such a problem all the history of the solar system could be evolved.

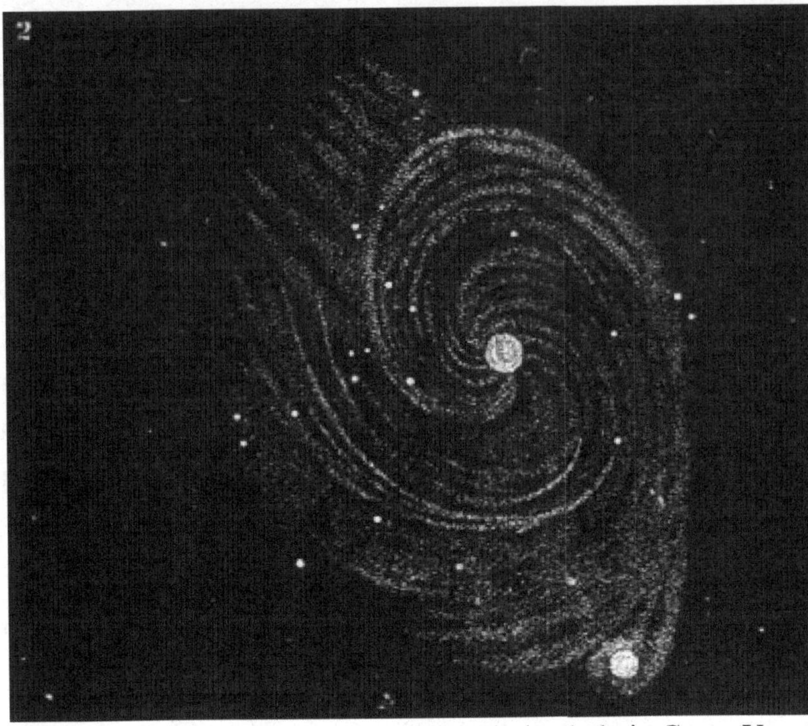

FIG. 80.—Lord Rosse's drawing of the spiral nebula in Canes Venatici, with the stub marks of the draughtsman unduly emphasised into features by the engraver.

Laplace pictured to himself this mass shrinking and thereby whirling more and more rapidly. A spinning body shrinking in size and retaining its original amount of rotation, as it will unless a brake is applied, must spin more and more rapidly as it shrinks. It has what mathematicians call a constant moment of momentum; and what it loses in leverage, as it shrinks, it gains in speed. The mass is held together by gravitation, every particle attracting every other particle; but since all the particles are describing curved paths, they will tend to fly off tangentially, and only a small excess of the gravitation force over the centrifugal is left to pull the particles in, and slowly to concentrate the nebula. The mutual gravitation of the parts is opposed by the centrifugal force of the whirl. At length a point is reached where the two forces balance. A portion outside a certain line will be in equilibrium; it will be left behind, and the rest must contract without it. A ring is formed, and away goes the inner nucleus contracting further and further towards a centre. After a time another ring will be left behind in the same way, and so on. What happens to these rings? They rotate with the motion they possess when thrown or shrunk off; but will they remain rings? If perfectly regular they may; if there be any irregularity they are liable to break up. They will break into one or two or more large masses, which are ultimately very likely to collide and become one. The revolving body so formed is still a rotating gaseous mass; and it will go on shrinking and cooling and throwing off rings, like the larger nucleus by which it has been abandoned. As any nucleus gets smaller, its rate of rotation increases, and so the rings last thrown off will be spinning faster than those thrown off earliest. The final nucleus or residual central body will be rotating fastest of all.

The nucleus of the whole original mass we now see shrunk up into what we call the sun, which is spinning on its axis once every twenty-five days. The rings successively thrown off by it are now the planets—some large, some small—those last thrown off rotating round him comparatively quickly, those outside much more slowly. The rings thrown off by the planetary gaseous masses as they contracted have now become satellites; except one ring which has remained without breaking up, and is to be seen rotating round Saturn still.

One other similar ring, an abortive attempt at a planet, is also left round the sun (the zone of asteroids).

Such, crudely and baldly, is the famous nebular hypothesis of Laplace. It was first stated, as has been said above, by the philosopher Kant, but it was elaborated into much fuller detail by the greatest of French mathematicians and astronomers.

The contracting masses will condense and generate great quantities of heat by their own shrinkage; they will at a certain stage condense to liquid, and after a time will begin to cool and congeal with a superficial crust, which will get thicker and thicker; but for ages they will remain hot, even after they have become thoroughly solid. The small ones will cool fastest; the big ones will retain their heat for an immense time. Bullets cool quickly, cannon-balls take hours or days to cool, planets take millions of years. Our moon may be nearly cold, but the earth is still warm—indeed, very hot inside. Jupiter is believed by some observers still to glow with a dull red heat; and the high temperature of the much larger and still liquid mass of the sun is apparent to everybody. Not till it begins to scum over will it be perceptibly cooler.

FIG. 81.—Saturn.

Many things are now known concerning heat which were not known to Laplace (in the above paragraph they are only hinted at), and these confirm and strengthen the general features of his hypothesis in a striking way; so do the most recent telescopic discoveries. But fresh possibilities have now occurred to us, tidal phenomena are seen to have an influence then wholly unsuspected, and it will be in a modified and amplified form that the philosopher of next century will still hold to the main features of this famous old Nebular Hypothesis respecting the origin of the sun and planets—the Evolution of the solar system.

NOTES TO LECTURE XII

The subject of stellar astronomy was first opened up by Sir William Herschel, the greatest observing astronomer.

Frederick William Herschel was born in Hanover in 1738, and brought up as a musician. Came to England in 1756. First saw a telescope in 1773. Made a great many himself, and began a survey of the heavens. His sister Caroline, born in 1750, came to England in 1772, and became his devoted assistant to the end of his life. Uranus discovered in 1781. Music finally abandoned next year, and the 40-foot telescope begun. Discovered two moons of Saturn and two of Uranus. Reviewed, described, and gauged all the visible heavens. Discovered and catalogued 2,500 nebulæ and 806 double stars. Speculated concerning the Milky Way, the nebulosity of stars, the origin and growth of solar systems. Discovered that the stars were in motion, not fixed, and that the sun as one of them was journeying towards a point in the constellation Hercules. Died in 1822, eighty-four years old. Caroline Herschel discovered eight comets, and lived on to the age of ninety-eight.

LECTURE XII

HERSCHEL AND THE MOTION OF THE FIXED STARS

WE may admit, I think, that, with a few notable exceptions, the work of the great men we have been recently considering was rather to complete and round off the work of Newton, than to strike out new and original lines.

This was the whole tendency of eighteenth century astronomy. It appeared to be getting into an adult and uninteresting stage, wherein everything could be calculated and predicted. Labour and ingenuity, and a severe mathematical training, were necessary to work out the remote consequences of known laws, but nothing fresh seemed likely to turn up. Consequently men's minds began turning in other directions, and we find chemistry and optics largely studied by some of the greatest minds, instead of astronomy.

But before the century closed there was destined to arise one remarkable exception—a man who was comparatively ignorant of that which had been done before—a man unversed in mathematics and the intricacies of science, but who possessed such a real and genuine enthusiasm and love of Nature that he overcame the force of adverse circumstances, and entering the territory of astronomy by a by-path, struck out a new line for himself, and infused into the science a healthy spirit of fresh life and activity.

This man was William Herschel.

"The rise of Herschel," says Miss Clerke, "is the one conspicuous anomaly in the otherwise somewhat quiet and prosy eighteenth century. It proved decisive of the course of events in the nineteenth. It was unexplained by anything that had gone before, yet all that came after hinged upon it. It gave a new direction to effort; it lent a fresh impulse to thought. It opened a channel for the widespread public interest which was gathering towards astronomical subjects to flow in."

Herschel was born at Hanover in 1738, the son of an oboe player in a military regiment. The father was a good musician, and a cultivated man. The mother was a German *Frau* of the period, a strong, active, business-like woman, of strong character and profound ignorance. Herself unable to write, she set her face against learning and all new-fangled notions. The education of the sons she could not altogether control, though she lamented over it, but the education of her two daughters she strictly limited to cooking, sewing, and household management. These, however, she taught them well.

It was a large family, and William was the fourth child. We need only remember the names of his younger brother Alexander, and of his much younger sister Caroline.

They were all very musical—the youngest boy was once raised upon a table to play the violin at a public performance. The girls were forbidden to learn music by their mother, but their father sometimes taught them a little on the sly. Alexander was besides an ingenious mechanician.

At the age of seventeen, William became oboist to the Hanoverian Guards, shortly before the regiment was ordered to England. Two years later he removed himself from the regiment, with the approval of his parents, though probably without the approbation or consent of the commanding officer, by whom such removal would be regarded as simple desertion, which indeed it was; and George III. long afterwards handed him an official pardon for it.

At the age of nineteen, he was thus launched in England with an outfit of some French, Latin, and English, picked up by himself; some skill in playing the hautboy, the violin, and the organ, as taught by his father; and some good linen and clothing, and an immense stock of energy, provided by his mother.

He lived as musical instructor to one or two militia bands in Yorkshire, and for three years we hear no more than this of him. But, at the end of that time, a noted organist, Dr. Miller, of Durham, who had heard his playing, proposed that he should come and live with him and play at concerts, which he was very glad to do. He next obtained the post of organist at Halifax; and some four or five years later he was invited to become organist at the Octagon Chapel in Bath, and soon led the musical life of that then very fashionable place.

About this time he went on a short visit to his family at Hanover, by all of whom he was very much beloved, especially by his young sister Caroline, who always regarded him as specially her own brother. It is rather pitiful, however, to find that her domestic occupations still unfairly repressed and blighted her life. She says:—

"Of the joys and pleasures which all felt at this long-wished-for meeting with my—let me say my dearest—brother, but a small portion could fall to my share; for with my constant attendance at church and school, besides the time I was employed in doing the drudgery of the scullery, it was but seldom I could make one in the group when the family were assembled together."

While at Bath he wrote many musical pieces—glees, anthems, chants, pieces for the harp, and an orchestral symphony. He taught a large number of pupils, and lived a hard and successful life. After fourteen hours or so spent in teaching and playing, he would retire at night to instruct his mind with a study of mathematics, optics, Italian, or Greek, in all of which he managed to make some progress. He also about this time fell in with some book on astronomy.

In 1763 his father was struck with paralysis, and two years later he died.

William then proposed that Alexander should come over from Hanover and join him at Bath, which was done. Next they wanted to rescue their sister Caroline from her humdrum existence, but this was a more difficult matter. Caroline's journal gives an account of her life at this time that is instructive. Here are a few extracts from it:—

"My father wished to give me something like a polished education, but my mother was particularly determined that it should be a rough, but at the same time a useful one; and nothing further she thought was necessary but to send me two or three months to a sempstress to be taught to make household linen....

"My mother would not consent to my being taught French, ... so all my father could do for me was to indulge me (and please himself) sometimes with a short lesson on the violin, when my mother was either in good humour or out of the way.... She had cause for wishing me not to know more than was necessary for being useful in the family; for it was her certain belief that my brother William would have returned to his country, and my eldest brother not have looked so high, if they had had a little less learning."

However, seven years after the death of their father, William went over to Germany and returned to England in triumph, bringing Caroline with him: she being then twenty-two.

So now began a busy life in Bath. For Caroline the work must have been tremendous. For, besides having to learn singing, she had to learn English. She had, moreover, to keep accounts and do the marketing.

When the season at Bath was over, she hoped to get rather more of her brother William's society; but he was deep in optics and astronomy, used to sleep with the books under his pillow, read them during meals, and scarcely ever thought of anything else.

He was determined to see for himself all the astronomical wonders; and there being a small Gregorian reflector in one of the shops, he hired it. But he was not satisfied with this, and contemplated making a telescope 20 feet long. He wrote to opticians inquiring the price of a mirror suitable, but found there were none so large, and that even the smaller ones were beyond his means. Nothing daunted, he determined to make some for himself. Alexander entered into his plans: tools, hones, polishers, and all sorts of rubbish were imported into the house, to the sister's dismay, who says:—

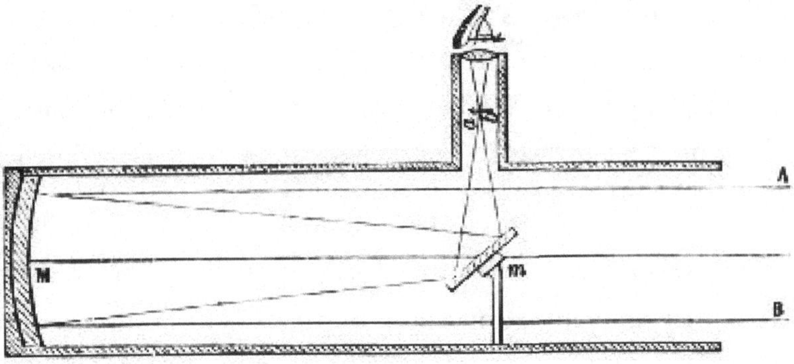

FIG. 82.—Principle of Newtonian reflector.

"And then, to my sorrow, I saw almost every room turned into a workshop. A cabinet-maker making a tube and stands of all descriptions in a handsomely furnished drawing-room; Alex. putting up a huge turning-machine (which he had brought in the autumn from Bristol, where he used to spend the summer) in a bed-room, for turning patterns, grinding glasses, and turning eye-pieces, &c. At the same time music durst not lie entirely dormant during the summer, and my brother had frequent rehearsals at home."

Finally, in 1774, at the age of thirty-six, he had made himself a 5½-foot telescope, and began to view the heavens. So attached was he to the instrument that he would run from the concert-room between the parts, and take a look at the stars.

He soon began another telescope, and then another. He must have made some dozen different telescopes, always trying to get them bigger and bigger; at last he got a 7-foot and then a 10-foot instrument, and began a systematic survey of the heavens; he also began to communicate his results to the Royal Society.

He now took a larger house, with more room for workshops, and a grass plot for a 20-foot telescope, and still he went on grinding mirrors—literally hundreds of them.

I read another extract from the diary of his sister, who waited on him and obeyed him like a spaniel:—

"My time was taken up with copying music and practising, besides attendance on my brother when polishing, since by way of keeping him alive I was constantly obliged to feed him by putting the victuals by bits into his mouth. This was once the case when, in order to finish a 7-foot mirror, he had not taken his hands from it for sixteen hours together. In general he was never unemployed at meals, but was always at those times contriving or making drawings of whatever came in his mind. Generally I was obliged to read to him whilst he was at the turning-lathe, or polishing mirrors—*Don Quixote*, *Arabian Nights' Entertainments*, the novels of Sterne, Fielding, &c.; serving tea and supper without interrupting the work with which he was engaged, ... and sometimes lending a hand. I became, in time, as useful a member of the workshop as a boy might be to his master in the first year of his apprenticeship.... But as I was to take a part the next year in the oratorios, I had, for a whole twelvemonth, two lessons per week from Miss Fleming, the celebrated dancing-mistress, to drill me for a gentlewoman (God knows how she succeeded). So we lived on without interruption. My brother Alex. was absent from Bath for some months every summer, but when at home he took much pleasure in executing some turning or clockmaker's work for his brother."

The music, and the astronomy, and the making of telescopes, all went on together, each at high pressure, and enough done in each to satisfy any ordinary activity. But the Herschels knew no rest. Grinding mirrors by day, concerts and oratorios in the evening, star-gazing at night. It is strange his health could stand it.

The star-gazing, moreover, was no *dilettante* work; it was based on a serious system—a well thought out plan of observation. It was nothing less than this—to pass the whole heavens steadily and in order through the telescope, noting and describing and recording every object that should be visible, whether previously known or unknown. The operation is called sweeping; but it is not a rapid passage from one object to another, as the term might suggest; it is a most tedious business, and consists in following with the telescope a certain field of view for some minutes, so as to be sure that nothing is missed, then shifting it to the next overlapping field, and watching again. And whatever object appears must be scrutinized anxiously to see what there is peculiar about it. If a star, it may be double, or it may be coloured, or it may be nebulous; or again it may be variable, and so its brightness must be estimated in order to compare with a subsequent observation.

Four distinct times in his life did Herschel thus pass the whole visible heavens under review; and each survey occupied him several years. He discovered double stars, variable stars, nebulæ, and comets; and Mr. William Herschel, of Bath, the amateur astronomer, was gradually emerging from his obscurity, and becoming a known man.

Tuesday, the 13th of March, 1781, is a date memorable in the annals of astronomy. "On this night," he writes to the Royal Society, "in examining the small stars near η Geminorum, I perceived one visibly larger than the rest. Struck with its uncommon appearance, I compared it to η Geminorum and another star, and finding it so much larger than either, I suspected it to be a comet."

The "comet" was immediately observed by professional astronomers, and its orbit was computed by some of them. It was thus found to move in nearly a circle instead of an elongated ellipse, and to be nearly twice as far from the sun as Saturn. It was no comet, it was a new planet; more than 100 times as big as the earth, and nearly twice as far away as Saturn. It was presently christened "Uranus."

This was a most striking discovery, and the news sped over Europe. To understand the interest it excited we must remember that such a discovery was unique. Since the most ancient times of which men had any knowledge, the planets Mercury, Venus, Mars, Jupiter, Saturn, had been known, and there had been no addition to their number. Galileo and others had discovered satellites indeed, but a new primary planet was an entire and utterly unsuspected novelty.

One of the most immediate consequences of the event was the discovery of Herschel himself. The Royal Society made him a Fellow the same year. The University of Oxford dubbed him a doctor; and the King sent for him to bring his telescope and show it at Court. So to London and Windsor he went, taking with him his best telescope. Maskelyne, the then Astronomer-Royal, compared it with the National one at Greenwich, and found Herschel's home-made instrument far the better of the two. He had a stand made after Herschel's pattern, but was so disgusted with his own instrument now that he scarcely thought it worthy of the stand when it was made. At Windsor, George III. was very civil, and Mr. Herschel was in great request to show the ladies of the Court Saturn and other objects of interest. Mr. Herschel exhibited a piece of worldly wisdom under these circumstances, that recalls faintly the behaviour of Tycho Brahé under similar circumstances. The evening when the exhibition was to take place threatened to become cloudy and wet, so Herschel rigged up an artificial Saturn, constructed of card and tissue paper, with a lamp behind it, in the distant wall of a garden; and, when the time came, his new titled friends were regaled with a view of this imitation Saturn through the telescope—the real one not being visible. They went away much pleased.

He stayed hovering between Windsor and Greenwich, and uncertain what was to be the outcome of all this regal patronizing. He writes to his sister that he would much rather be back grinding mirrors at Bath. And she writes begging him to come, for his musical pupils were getting impatient. They had to get the better of their impatience, however, for the King ultimately appointed him astronomer or rather telescope-maker to himself, and so Caroline and the whole household were sent for, and established in a small house at Datchet.

From being a star-gazing musician, Herschel thus became a practical astronomer. Henceforth he lived in his observatory; only on wet and moonlight nights could he be torn away from it. The day-time he devoted to making his long-contemplated 20-foot telescope.

Not yet, however, were all their difficulties removed. The house at Datchet was a tumble-down barn of a place, chosen rather as a workshop and observatory than as a dwelling-house. And the salary allowed him by George III. was scarcely a princely one. It was, as a matter of fact, £200 a year. The idea was that he would earn his living by making telescopes, and so indeed he did. He made altogether some hundreds. Among others, four for the King. But this eternal making of telescopes for other people to use or play with was a weariness to the flesh. What he wanted was to observe, observe, observe.

Sir William Watson, an old friend of his, and of some influence at Court, expressed his mind pretty plainly concerning Herschel's position; and as soon as the King got to understand that there was anything the matter, he immediately offered £2,000 for a gigantic telescope to be made for Herschel's own use. Nothing better did he want in life. The whole army of carpenters and craftsmen resident in Datchet were pressed into the service. Furnaces for the speculum metal were built, stands erected, and the 40-foot telescope fairly begun. It cost £4,000 before it was finished, but the King paid the whole.

FIG. 83.—Herschel's 40-foot telescope.

With it he discovered two more satellites to Saturn (five hitherto had been known), and two moons to his own planet Uranus. These two are now known as Oberon and Titania. They were not seen again till some forty years after, when his son, Sir John Herschel, reobserved them. And in 1847, Mr. Lassell, at his house, "Starfield," near Liverpool, discovered two more, called Ariel and Umbriel, making the number four, as now known. Mr. Lassell also discovered, with a telescope of his own making, an eighth satellite of Saturn—Hyperion—and a satellite to Neptune.

A letter from a foreign astronomer about this period describes Herschel and his sister's method of work:—

"I spent the night of the 6th of January at Herschel's, in Datchet, near Windsor, and had the good luck to hit on a fine evening. He has his 20-foot Newtonian telescope in the open air, and mounted in his garden very simply and conveniently. It is moved by an assistant, who stands below it.... Near the instrument is a clock regulated to sidereal time.... In the room near it sits Herschel's sister, and she has Flamsteed's atlas open before her. As he gives her the word, she writes down the declination and right ascension, and the other circumstances of the observation. In this way Herschel examines the whole sky without omitting the least part. He commonly observes with a magnifying power of one hundred and fifty, and is sure that after four or five years he will have passed in review every object above our horizon. He showed me the book in which his observations up to this time are written, and I am astonished at the great number of them. Each sweep covers 2° 15' in declination, and he lets each star pass at least three times through the field of his telescope, so that it is impossible that anything can escape him. He has already found about 900 double stars, and almost as many nebulæ. I went to bed about one o'clock, and up to that time he had found that night four or five new nebulæ. The thermometer in the garden stood at 13° Fahrenheit; but, in spite of this, Herschel observes the whole night through, except that he stops every three or four hours and goes into the room for a few moments. For some years Herschel has observed the heavens every hour when the weather is clear, and this always in the open air, because he says that the telescope only performs well when it is at the same temperature as the air. He protects himself against the weather by putting on more clothing. He has an excellent constitution, and thinks about nothing else in the world but the celestial bodies. He has promised me in the most cordial way, entirely in the service of astronomy, and without thinking of his own interest, to see to the telescopes I have ordered for European observatories, and he will himself attend to the preparation of the mirrors."

FIG. 84.—William Herschel.
From an Original Picture in the Possession of WM. WATSON, M.D., F.R.S.
Painted by Abbott. Engraved by Ryder.

In 1783, Herschel married an estimable lady who sympathized with his pursuits. She was the only daughter of a City magnate, so his pecuniary difficulties, such as they were (they were never very troublesome to him), came to an end. They moved now into a more commodious house at Slough. Their one son, afterwards the famous Sir John Herschel, was born some nine years later. But the marriage was rather a blow to his devoted sister: henceforth she lived in lodgings, and went over at night-time to help him observe. For it must be remarked that this family literally turned night into day. Whatever sleep they got was in the day-time. Every fine night without exception was spent in observing: and the quite incredible fierceness of the pursuit is illustrated, as strongly as it can be, by the following sentence out of Caroline's diary, at the time of the move from Datchet to Slough: "The last night at Datchet was spent in sweeping till daylight, and by the next evening the telescope stood ready for observation at Slough."

Caroline was now often allowed to sweep with a small telescope on her own account. In this way she picked up a good many nebulæ in the course of her life, and eight comets, four of which were quite new, and one of which, known since as Encke's comet, has become very famous.

The work they got through between them is something astonishing. He made with his own hands 430 parabolic mirrors for reflecting telescopes, besides a great number of complete instruments. He was forty-two when he began contributing to the Royal Society; yet before he died he had sent them sixty-nine long and elaborate treatises. One of these memoirs is a catalogue of 1000 nebulæ. Fifteen years after he sends in another 1000; and some years later another 500. He also discovered 806 double stars, which he proved were really corrected from the fact that they revolved round each other (p. 309). He lived to see some of them perform half a revolution. For him the stars were not fixed: they moved slowly among themselves. He detected their proper motions. He passed the whole northern firmament in review four distinct times; counted the stars in 3,400 gauge-fields, and estimated the brightness of hundreds of stars. He also measured as accurately as he could their proper motions, devising for this purpose the method which still to this day remains in use.

And what is the outcome of it all? It is not Uranus, nor the satellites, nor even the double stars and the nebulæ considered as mere objects: it is the beginning of a science of the stars.

FIG. 85.—Caroline Herschel.
From a Drawing from Life, by GEORGE MÜLLER, 1847.

Hitherto the stars had only been observed for nautical and practical purposes. Their times of rising and southing and setting had been noted; they had been treated as a clock or piece of dead mechanism, and as fixed points of reference. All the energies of astronomers had gone out towards the solar system. It was the planets that had been observed. Tycho had observed and tabulated their positions. Kepler had found out some laws of their motion.

Galileo had discovered their peculiarities and attendants. Newton and Laplace had perceived every detail of their laws.

But for the stars—the old Ptolemaic system might still have been true. They might still be mere dots in a vast crystalline sphere, all set at about one distance, and subservient to the uses of the earth.

Herschel changed all this. Instead of sameness, he found variety; instead of uniformity of distance, limitless and utterly limitless fields and boundless distances; instead of rest and quiescence, motion and activity; instead of stagnation, life.

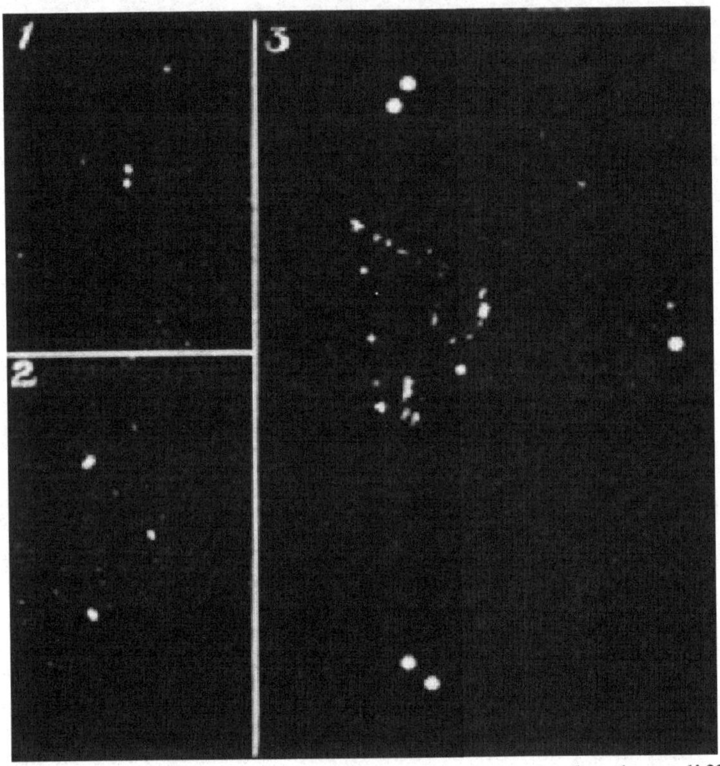

FIG. 86.—The double-double star ε Lyræ as seen under three different powers.

Yes, that is what Herschel discovered—the life and activity of the whole visible universe. No longer was our little solar system to be the one object of regard, no longer were its phenomena to be alone interesting to man. With Herschel every star was a solar system. And more than that: he found suns revolving round suns, at distances such as the mind reels at, still obeying the same law of gravitation as pulls an apple from a tree. He tried hard to estimate the distance of the stars from the earth, but there he failed: it was too hopeless a problem. It was solved some time after his death by Bessel, and the distances of many stars are now known but these distances are awful and unspeakable. Our distance from the sun shrinks up into a mere speck—the whole solar system into a mere unit of measurement, to be repeated hundreds of thousands of times before we reach the stars.

Yet their motion is visible—yes, to very accurate measurement quite plain. One star, known as 61 Cygni, was then and is now rushing along at the rate of 100 miles every second. Not that you must imagine that this makes any obvious and apparent change in its position. No, for all ordinary and practical purposes they are still fixed stars; thousands of years will show us no obvious change; "Adam" saw precisely the same constellations as we do: it is only by refined micrometric measurement with high magnifying power that their flight can be detected.

But the sun is one of the stars—not by any means a specially large or bright one; Sirius we now know to be twenty times as big as the sun. The sun is one of the stars: then is it at rest? Herschel asked this question and endeavoured to answer it. He succeeded in the most astonishing manner. It is, perhaps, his most remarkable discovery, and savours of intuition. This is how it happened. With imperfect optical means and his own eyesight to guide him, he considered and pondered over the proper motion of the stars as he had observed it, till he discovered a kind of uniformity running through it all. Mixed up with irregularities and individualities, he found that in a certain part of the heavens the stars were on the whole opening out—separating slowly from each other; on the opposite side of the heavens they were on the average closing up—getting slightly nearer to each other; while in directions at right angles to this they were fairly preserving their customary distances asunder.

Now, what is the moral to be drawn from such uniformity of behaviour among unconnected bodies? Surely that this part of their motion is only apparent—that it is we who are moving. Travelling over a prairie bounded by a belt of trees, we should see the trees in our line of advance opening out, and those behind closing up; we should see in fact the same kind of apparent motion as Herschel was able to detect among the stars: the opening out being most marked near the constellation Hercules. The conclusion is obvious: the sun, with all its planets, must be steadily moving towards a point in the constellation Hercules. The most accurate modern research has been hardly able to improve upon this statement of Herschel's. Possibly the solar system may ultimately be found to revolve round some other body, but what that is no one knows. All one can tell is the present direction of the majestic motion: since it was discovered it has continued unchanged, and will probably so continue for thousands of years.

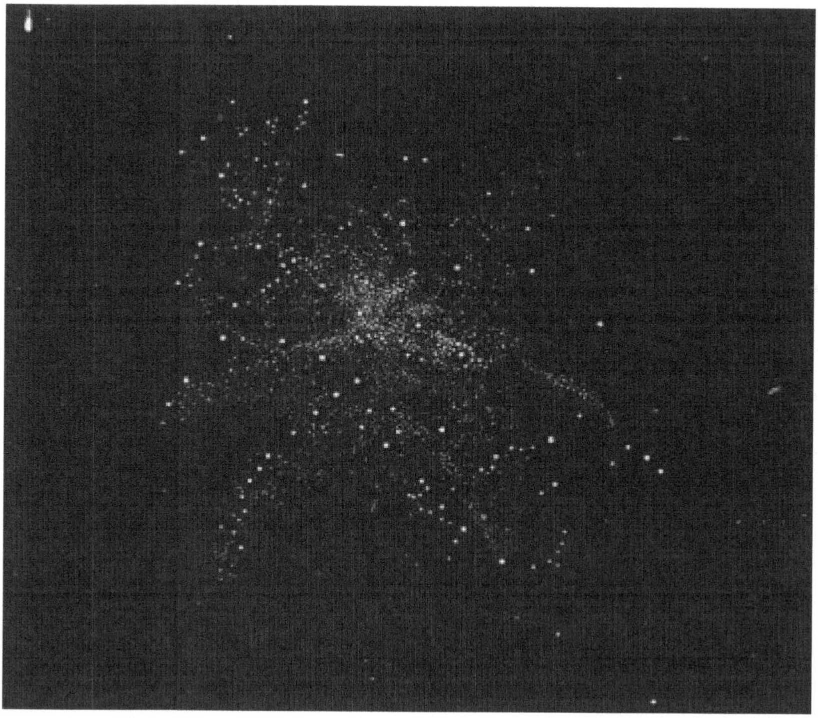

FIG. 87.—Old drawing of the cluster in Hercules.

And, finally, concerning the nebulæ. These mysterious objects exercised a strong fascination for Herschel, and many are the speculations he indulges in concerning them. At one time he regards them all as clusters of stars, and the Milky Way as our cluster; the others he regards as other universes almost infinitely distant; and he proceeds to gauge and estimate the shape of our own universe or galaxy of suns, the Milky Way.

Later on, however, he pictures to himself the nebulæ as nascent suns: solar systems before they are formed. Some he thinks have begun to aggregate, while some are still glowing gas.

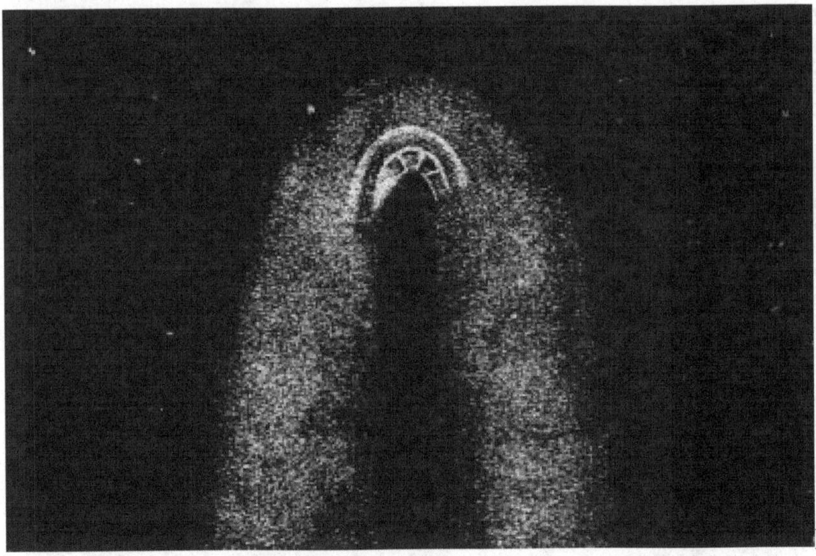

FIG. 88.—Old drawing of the Andromeda nebula.

He likens the heavens to a garden in which there are plants growing in all manner of different stages: some shooting, some in leaf, some in flower, some bearing seed, some decaying; and thus at one inspection we have before us the whole life-history of the plant.

Just so he thinks the heavens contain worlds, some old, some dead, some young and vigorous, and some in the act of being formed. The nebulæ are these latter, and the nebulous stars are a further stage in the condensation towards a sun.

And thus, by simple observation, he is led towards something very like the nebular hypothesis of Laplace; and his position, whether it be true or false, is substantially the same as is held to-day.

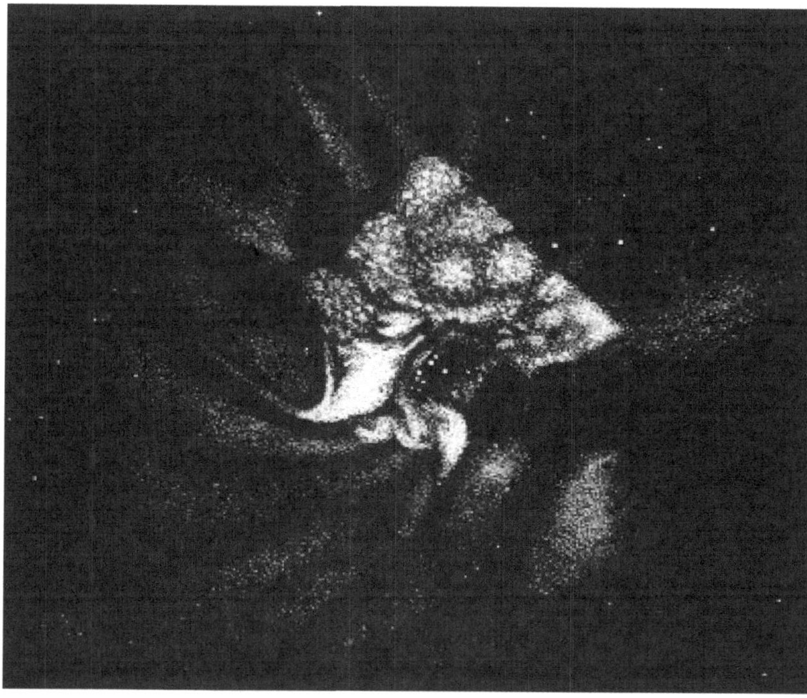

FIG. 89.—The great nebula in Orion.

We *know* now that many of the nebulæ consist of innumerable isolated particles and may be spoken of as gas. We know that some are in a state of whirling motion. We know also that such gas left to itself will slowly as it cools condense and shrink, so as to form a central solid nucleus; and also, if it were in whirling motion, that it would send off rings from itself, and that these rings could break up into planets. In two familiar cases the ring has not yet thus aggregated into planet or satellite—the zone of asteroids, and Saturn's ring.

The whole of this could not have been asserted in Herschel's time: for further information the world had to wait.

These are the problems of modern astronomy—these and many others, which are the growth of this century, aye, and the growth of the last thirty or forty, and indeed of the last ten years. Even as I write, new and very confirmatory discoveries are being announced. The Milky Way *does* seem to have some affinity with our sun. And the chief stars of the constellation of Orion constitute another family, and are enveloped in the great nebula, now by photography perceived to be far greater than had ever been imagined.

What is to be the outcome of it all I know not; but sure I am of this, that the largest views of the universe that we are able to frame, and the grandest manner of its construction that we can conceive, are certain to pale and shrink and become inadequate when confronted with the truth.

NOTES TO LECTURE XIII

BODE'S LAW.—Write down the series 0, 3, 6, 12, 24, 48, &c.; add 4 to each, and divide by 10; you get the series:

·4	·7	1·0	1·6	2·8	5·2	10·0	19·6	38·8
Mercury	Venus	Earth	Mars	—	Jupiter	Saturn	Uranus	—

numbers which very fairly represent the distances of the then known planets from the sun in the order specified.

Ceres was discovered on the 1st of January, 1801, by Piazzi; Pallas in March, 1802, by Olbers; Juno in 1804, by Harding; and

Vesta in 1807, by Olbers. No more asteroids were discovered till 1845, but there are now several hundreds known. Their diameters range from 500 to 20 miles.

Neptune was discovered from the perturbations of Uranus by sheer calculation, carried on simultaneously and independently by Leverrier in Paris, and Adams in Cambridge. It was first knowingly seen by Galle, of Berlin, on the 23rd of September, 1846.

LECTURE XIII

THE DISCOVERY OF THE ASTEROIDS

UP to the time of Herschel, astronomical interest centred on the solar system. Since that time it has been divided, and a great part of our attention has been given to the more distant celestial bodies. The solar system has by no means lost its interest—it has indeed gained in interest continually, as we gain in knowledge concerning it; but in order to follow the course of science it will be necessary for us to oscillate to and fro, sometimes attending to the solar system—the planets and their satellites—sometimes extending our vision to the enormously more distant stellar spaces.

Those who have read the third lecture in Part I. will remember the speculation in which Kepler indulged respecting the arrangements of the planets, the order in which they succeeded one another in space, and the law of their respective distances from the sun; and his fanciful guess about the five regular solids inscribed and circumscribed about their orbits.

The rude coincidences were, however, accidental, and he failed to discover any true law. No thoroughly satisfactory law is known at the present day. And yet, if the nebular hypothesis or anything like it be true, there must be some law to be discovered hereafter, though it may be a very complicated one.

An empirical relation is, however, known: it was suggested by Tatius, and published by Bode, of Berlin, in 1772. It is always known as Bode's law.

Bode's law asserts that the distance of each planet is approximately double the distance of the inner adjacent planet from the sun, but that the rate of increase is distinctly slower than this for the inner ones; consequently a better approximation will be obtained by adding a constant to each term of an appropriate geometrical progression. Thus, form a doubling series like this, 1½, 3, 6, 12, 24, &c. doubling each time; then add 4 to each, and you get a series which expresses very fairly the relative distances of the successive planets from the sun, except that the number for Mercury is rather erroneous, and we now know that at the other extreme the number for Neptune is erroneous too.

I have stated it in the notes above in a form calculated to give the law every chance, and a form that was probably fashionable after the discovery of Uranus; but to call the first term of the doubling series 0 is evidently not quite fair, though it puts Mercury's distance right. Neptune's distance, however, turns out to be more nearly 30 times the earth's distance than 38·8. The others are very nearly right: compare column D of the table preceding Lecture III. on p. 57, with the numbers in the notes on p. 294.

The discovery of Uranus a few years afterwards, in 1781, at 19·2 times the earth's distance from the sun, lent great *éclât* to the law, and seemed to establish its right to be regarded as at least a close approximation to the truth.

The gap between Mars and Jupiter, which had often been noticed, and which Kepler filled with a hypothetical planet too small to see, comes into great prominence by this law of Bode. So much so, that towards the end of last century an enthusiastic German, von

Zach, after some search himself for the expected planet, arranged a committee of observing astronomers, or, as he termed it, a body of astronomical detective police, to begin a systematic search for this missing subject of the sun.

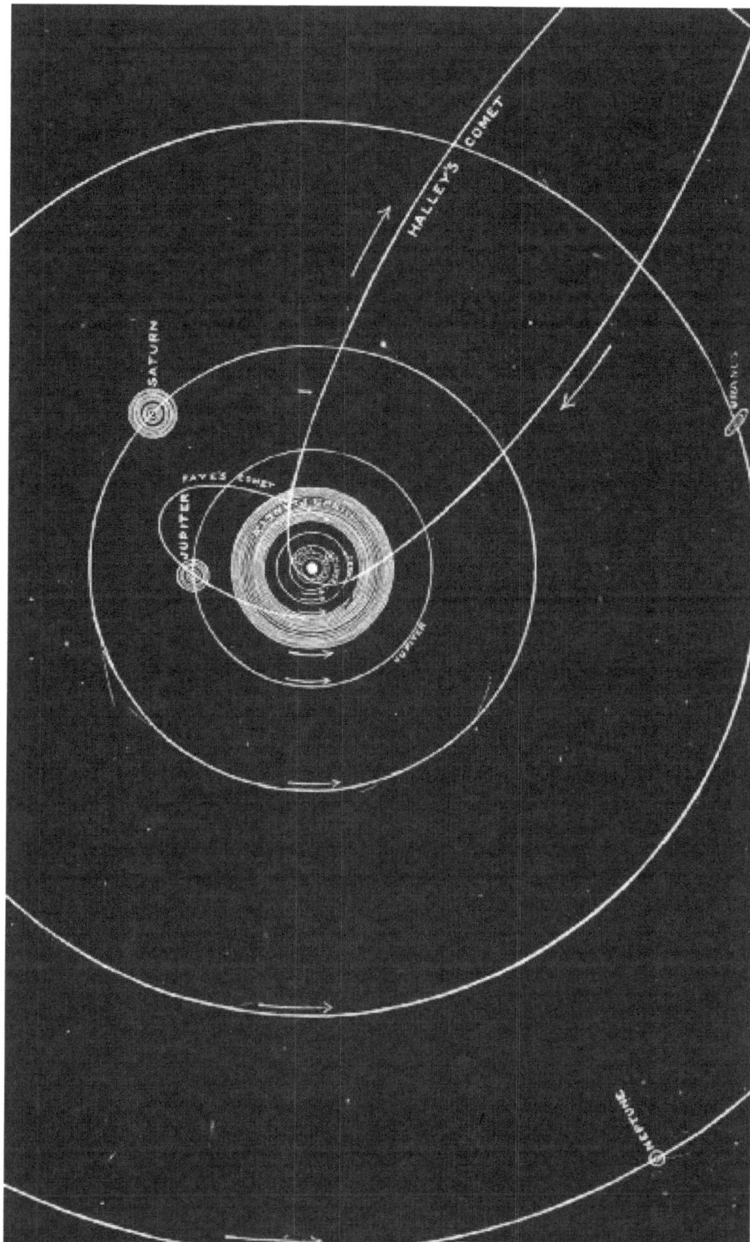

FIG. 90.—Planetary orbits to scale; showing the Asteroidal region between Jupiter and Mars. (The orbits of satellites are exaggerated.)

In 1800 the preliminaries were settled: the heavens near the zodiac were divided into twenty-four regions, each of which was intrusted to one observer to be swept. Meanwhile, however, quite independently of these arrangements in Germany, and entirely unknown to this committee, a quiet astronomer in Sicily, Piazzi, was engaged in making a catalogue of the stars. His attention was directed to a certain region in Taurus by an error in a previous catalogue, which contained a star really non-existent.

In the course of his scrutiny, on the 1st of January, 1801, he noticed a small star which next evening appeared to have shifted. He watched it anxiously for successive evenings, and by the 24th of January he was quite sure he had got hold of some moving body, not a star: probably, he thought, a comet. It was very small, only of the eighth magnitude; and he wrote to two astronomers (one of them Bode himself) saying what he had observed. He continued to observe till the 11th of February, when he was attacked by illness and compelled to cease.

His letters did not reach their destination till the end of March. Directly Bode opened his letter he jumped to the conclusion that this must be the missing planet. But unfortunately he was unable to verify the guess, for the object, whatever it was, had now got too near the sun to be seen. It would not be likely to be out again before September, and by that time it would be hopelessly lost again, and have just as much to be rediscovered as if it had never been seen.

Mathematical astronomers tried to calculate a possible orbit for the body from the observations of Piazzi, but the observed places were so desperately few and close together. It was like having to determine a curve from three points close together. Three observations ought to serve,[27] but if they are taken with insufficient interval between them it is extremely difficult to construct the whole circumstances of the orbit from them. All the calculations gave different results, and none were of the slightest use.

The difficulty as it turned out was most fortunate. It resulted in the discovery of one of the greatest mathematicians, perhaps the greatest, that Germany has ever produced—Gauss. He was then a young man of twenty-five, eking out a living by tuition. He had invented but not published several powerful mathematical methods (one of them now known as "the method of least squares"), and he applied them to Piazzi's observations. He was thus able to calculate an orbit, and to predict a place where, by the end of the year, the planet should be visible. On the 31st of December of that same year, very near the place predicted by Gauss, von Zach rediscovered it, and Olbers discovered it also the next evening. Piazzi called it Ceres, after the tutelary goddess of Sicily.

Its distance from the sun as determined by Gauss was 2·767 times the earth's distance. Bode's law made it 2·8. It was undoubtedly the missing planet. But it was only one hundred and fifty or two hundred miles in diameter—the smallest heavenly body known at the time of its discovery. It revolves the same way as other planets, but the plane of its orbit is tilted 10° to the plane of the ecliptic, which was an exceptionally large amount.

Very soon, a more surprising discovery followed. Olbers, while searching for Ceres, had carefully mapped the part of the heavens where it was expected; and in March, 1802, he saw in this place a star he had not previously noticed. In two hours he detected its motion, and in a month he sent his observations to Gauss, who returned as answer the calculated orbit. It was distant 2·67, like Ceres, and was a little smaller, but it had a very excentric orbit: its plane being tilted 34½°, an extraordinary inclination. This was called Pallas.

Olbers at once surmised that these two planets were fragments of a larger one, and kept an eager look out for other fragments.

In two years another was seen, in the course of charting the region of the heavens traversed by Ceres and Pallas. It was smaller than either, and was called Juno.

In 1807 the persevering search of Olbers resulted in the discovery of another, with a very oblique orbit, which Gauss named Vesta. Vesta is bigger than any of the others, being five hundred miles in diameter, and shines like a star of the sixth magnitude. Gauss by this time had become so practised in the difficult computations that he worked out the complete orbit of Vesta within ten hours of receiving the observational data from Olbers.

For many weary years Olbers kept up a patient and unremitting search for more of these small bodies, or fragments of the large planet as he thought them; but his patience went unrewarded, and he died in 1840 without seeing or knowing of any more. In 1845 another was found, however, in Germany, and a few weeks later two others by Mr. Hind in England. Since then there seems no end to them; numbers have been discovered in America, where Professors

Peters and Watson have made a specialty of them, and have themselves found something like a hundred.

Vesta is the largest—its area being about the same as that of Central Europe, without Russia or Spain—and the smallest known is about twenty miles in diameter, or with a surface about the size of Kent. The whole of them together do not nearly equal the earth in bulk.

The main interest of these bodies to us lies in the question, What is their history? Can they have been once a single planet broken up? or are they rather an abortive attempt at a planet never yet formed into one?

The question is not *entirely* settled, but I can tell you which way opinion strongly tends at the present time.

Imagine a shell travelling in an elliptic orbit round the earth to suddenly explode: the centre of gravity of all its fragments would continue moving along precisely the same path as had been traversed by the centre of the shell before explosion, and would complete its orbit quite undisturbed. Each fragment would describe an orbit of its own, because it would be affected by a different initial velocity; but every orbit would be a simple ellipse, and consequently every piece would in time return through its starting-point—viz. the place at which the explosion occurred. If the zone of asteroids had a common point through which they all successively passed, they could be unhesitatingly asserted to be the remains of an exploded planet. But they have nothing of the kind; their orbits are scattered within a certain broad zone—a zone everywhere as broad as the earth's distance from the sun, 92,000,000 miles—with no sort of law indicating an origin of this kind.

It must be admitted, however, that the fragments of our supposed shell might in the course of ages, if left to themselves, mutually perturb each other into a different arrangement of orbits from that with which they began. But their perturbations would be very minute, and moreover, on Laplace's theory, would only result in periodic changes, provided each mass were rigid. It is probable that the asteroids were at one time not rigid, and hence it is difficult to say what may have happened to them; but there is not the least reason to believe that their present arrangement is derivable in any way from an explosion, and it is certain that an enormous time must have elapsed since such an event if it ever occurred.

It is far more probable that they never constituted one body at all, but are the remains of a cloudy ring thrown off by the solar system in shrinking past that point: a small ring after the immense effort which produced Jupiter and his satellites: a ring which has aggregated into a multitude of little lumps instead of a few big ones. Such an event is not unique in the solar system; there is a similar ring round Saturn. At first sight, and to ordinary careful inspection, this differs from the zone of asteroids in being a solid lump of matter, like a quoit. But it is easy to show from the theory of gravitation, that a solid ring could not possibly be stable, but would before long get precipitated excentrically upon the body of the planet. Devices have been invented, such as artfully distributed irregularities calculated to act as satellites and maintain stability; but none of these things really work. Nor will it do to imagine the rings fluid; they too would destroy each other. The mechanical behaviour of a system of rings, on different hypotheses as to their constitution, has been worked out with consummate skill by Clerk Maxwell; who finds that the only possible constitution for Saturn's assemblage of rings is a multitude of discrete particles each pursuing its independent orbit. Saturn's ring is, in fact, a very concentrated zone of minor asteroids, and there is every reason to conclude that the origin of the solar asteroids cannot be very unlike the origin of the Saturnian ones. The nebular hypothesis lends itself readily to both.

The interlockings and motions of the particles in Saturn's rings are most beautiful, and have been worked out and stated by Maxwell with marvellous completeness. His paper constituted what is called "The Adams Prize Essay" for 1856. Sir George Airy, one of the adjudicators (recently Astronomer-Royal), characterized it as "one of the most remarkable applications of mathematics to physics that I have ever seen."

There are several distinct constituent rings in the entire Saturnian zone, and each perturbs the other, with the result that they ripple and pulse in concord. The waves thus formed absorb the effect of the mutual perturbations, and prevent an accumulation which would be dangerous to the persistence of the whole.

The only effect of gravitational perturbation and of collisions is gradually to broaden out the whole ring, enlarging its outer and diminishing its inner diameter. But if there were any frictional resistance in the medium through which the rings spin, then other effects would slowly occur, which ought to be looked for with interest. So complete and intimate is the way Maxwell works out and describes the whole circumstances of the motion of such an assemblage of particles, and so cogent his argument as to the necessity that they must move precisely so, and no otherwise, else the rings would not be stable, that it was a Cambridge joke concerning him that he paid a visit to Saturn one evening, and made his observations on the spot.

NOTES TO LECTURE XIV

The total number of stars in the heavens visible to a good eye is about 5,000. The total number at present seen by telescope is about 50,000,000. The number able to impress a photographic plate has not yet been estimated; but it is enormously greater still. Of those which we can see in these latitudes, about 14 are of the first magnitude, 48 of the second, 152 of the third, 313 of the fourth, 854 of the fifth, and 2,010 of the sixth; total, 3,391.

The quickest-moving stars known are a double star of the sixth magnitude, called 61 Cygni, and one of the seventh magnitude, called Groombridge 1830. The velocity of the latter is 200 miles a second. The nearest known stars are 61 Cygni and α Centauri. The distance of these from us is about 400,000 times the distance of the sun. Their parallax is accordingly half a second of arc. Sirius is more than a million times further from us than our sun is, and twenty times as big; many of the brightest stars are at more than double this distance. The distance of Arcturus is too great to measure even now. Stellar parallax was first securely detected in 1838, by Bessel, for

61 Cygni. Bessel was born in 1784, and died in 1846, shortly before the discovery of Neptune.

The stars are suns, and are most likely surrounded by planets. One planet belonging to Sirius has been discovered. It was predicted by Bessel, its position calculated by Peters, and seen by Alvan Clark in 1862. Another predicted one, belonging to Procyon, has not yet been seen.

A velocity of 5 miles a second could carry a projectile right round the earth. A velocity of 7 miles a second would carry it away from the earth, and round the sun. A velocity of 27 miles a second would carry a projectile right out of the solar system never to return.

LECTURE XIV

BESSEL—THE DISTANCES OF THE STARS, AND THE DISCOVERY OF STELLAR PLANETS

WE will now leave the solar system for a time, and hastily sketch the history of stellar astronomy from the time of Sir William Herschel.

You remember how greatly Herschel had changed the aspect of the heavens for man,—how he had found that none of the stars were really fixed, but were moving in all manner of ways: some of this motion only apparent, much of it real. Nevertheless, so enormously distant are they, that if we could be transported back to the days of the old Chaldæan astronomers, or to the days of Noah, we should still see the heavens with precisely the same aspect as they wear now. Only by refined apparatus could any change be discoverable in all those centuries. For all practical purposes, therefore, the stars may still be well called fixed.

Another thing one may notice, as showing their enormous distances, is that from every planet of the solar system the aspect of the heavens will be precisely the same. Inhabitants of Mars, or Jupiter, or Saturn, or Uranus, will see exactly the same constellations as we do. The whole dimensions of the solar system shrink up into a speck when so contemplated. And from the stars none of the planetary orbs of our system are visible at all; nothing but the sun is visible, and that merely as a twinkling star, brighter than some, but fainter than many others.

The sun and the stars are one. Try to realize this distinctly, and keep it in mind. I find it often difficult to drive this idea home. After some talk on the subject a friendly auditor will report, "the lecturer then described the stars, including that greatest and most magnificent of all stars, the sun." It would be difficult more completely to misapprehend the entire statement. When I say the sun is one of the stars, I mean one among the others; we are a long way from them, they are a long way from each other. They need be no more closely packed among each other than we are closely packed among them; except that some of them are double or multiple, and we are not double.

It is highly desirable to acquire an intimate knowledge of the constellations and a nodding acquaintance with their principal stars. A description of their peculiarities is dull and uninteresting unless they are at least familiar by name. A little *vivâ voce* help to begin with, supplemented by patient night scrutiny with a celestial globe or star maps under a tent or shed, is perhaps the easiest way: a very convenient instrument for the purpose of learning the constellations is the form of map called a "planisphere," because it can be made to show all the constellations visible at a given time at a given date, and no others. The Greek alphabet also is a thing that should be learnt by everybody. The increased difficulty in teaching science owing to the modern ignorance of even a smattering of Greek is becoming grotesque. The stars are named from their ancient grouping into constellations, and by the prefix of a Greek letter to the larger ones, and of numerals to the smaller ones. The biggest of all have special Arabic names as well. The brightest stars are called of "the first magnitude," the next are of "the second magnitude," and so on. But this arrangement into magnitudes has become technical and precise, and intermediate or fractional magnitudes are inserted. Those brighter than the ordinary first magnitude are therefore now spoken of as of magnitude ½, for instance, or ·6, which is rather confusing. Small telescopic stars are often only named by their numbers in some specified catalogue—a dull but sufficient method.

Here is a list of the stars visible from these latitudes, which are popularly considered as of the first magnitude. All of them should be familiarly recognized in the heavens, whenever seen.

Star.	Constellation.
Sirius	Canis major
Procyon	Canis minor
Rigel	Orion
Betelgeux	Orion
Castor	Gemini
Pollux	Gemini
Aldebaran	Taurus
Arcturus	Boötes
Vega	Lyra
Capella	Auriga
Regulus	Leo
Altair	Aquila
Fomalhaut	Southern Fish
Spica	Virgo

α Cygni is a little below the first magnitude. So, perhaps, is Castor. In the southern heavens, Canopus and α Centauri rank next after Sirius in brightness.

FIG. 91.—Diagram illustrating Parallax.

The distances of the fixed stars had, we know, been a perennial problem, and many had been the attempts to solve it. All the methods of any precision have depended on the Copernican fact that the earth in June was 184 million miles away from its position in December, and that accordingly the grouping and aspect of the heavens should be somewhat different when seen from so different a point of view. An apparent change of this sort is called generally parallax; *the* parallax of a star being technically defined as the angle subtended at the star by the radius of the earth's orbit: that is to say, the angle EσS; where E is the earth, S the sun, and σ a star (Fig. 91).

Plainly, the further off σ is, the more nearly parallel will the two lines to it become. And the difficulty of determining the parallax was just this, that the more accurately the observations were made, the more nearly parallel did those lines become. The angle was, in fact, just as likely to turn out negative as positive—an absurd result, of course, to be attributed to unavoidable very minute inaccuracies.

For a long time absolute methods of determining parallax were attempted; for instance, by observing the position of the star with respect to the zenith at different seasons of the year. And many of these determinations appeared to result in success. Hooke fancied he had measured a parallax for Vega in this way, amounting to 30" of arc. Flamsteed obtained 40" for γ Draconis. Roemer made a serious attempt by comparing observations of Vega and Sirius, stars almost the antipodes of each other in the celestial vault; hoping to detect some effect due to the size of the earth's orbit, which should apparently displace them with the season of the year. All these fancied results however, were shown to be spurious, and their real cause assigned, by the great discovery of the aberration of light by Bradley.

After this discovery it was possible to watch for still outstanding very minute discrepancies; and so the problem of stellar parallax was attacked with fresh vigour by Piazzi, by Brinkley, and by Struve. But when results were obtained, they were traced after long discussion to age and gradual wear of the instrument, or to some other minute inaccuracy. The more carefully the observation was made, the more nearly zero became the parallax—the more nearly infinite the distance of the stars. The brightest stars were the ones commonly chosen for the investigation, and Vega was a favourite, because, going near the zenith, it was far removed from the fluctuating and tiresome disturbances of atmospheric refraction. The reason bright stars were chosen was because they were presumably nearer than the others; and indeed a rough guess at their probable distance was made by supposing them to be of the same size as the sun, and estimating their light in comparison with sunlight. By this confessedly unsatisfactory method it had been estimated that Sirius must be 140,000 times further away than the sun is, if he be equally big. We now know that Sirius is much further off than this; and accordingly that he is much brighter, perhaps sixty times as bright, though not necessarily sixty times as big, as our sun. But even supposing him of the same light-giving power as the sun, his parallax was estimated as 1"·8, a quantity very difficult to be sure of in any absolute determination.

Relative methods were, however, also employed, and the advantages of one of these (which seems to have been suggested by Galileo) so impressed themselves upon William Herschel that he made a serious attempt to compass the problem by its means. The method was to take two stars in the same telescopic field and carefully to estimate their apparent angular distance from each other at different seasons of the year. All such disturbances as precession, aberration, nutation, refraction, and the like, would affect them both equally, and could thus be eliminated. If they were at the same distance from the solar system, relative parallax would, indeed, also be eliminated; but if, as was probable, they were at different distances, then they would apparently shift relatively to one another, and the amount of shift, if it could be observed, would measure, not indeed the distance of either from the earth, but their distance from each other. And this at any rate would be a step. It might be completed by similarly treating other stars in the same field, taking them in pairs together. A bright and a faint star would naturally be suitable, because their distances were likely to be unequal; and so Herschel fixed upon a number of doublets which he knew of, containing one bright and one faint component. For up to that time it had been supposed that such grouping in occasional pairs or triplets was chance coincidence, the two being optically foreshortened together, but having no real connection or proximity. Herschel failed in what he was looking for, but instead of that he discovered the real connection of a number of these doublets, for he found that they were slowly revolving round each other. There are a certain number of merely optical or accidental doublets, but the majority of them are real pairs of suns revolving round each other.

This relative method of mapping micrometrically a field of neighbouring stars, and comparing their configuration now and six months hence, was, however, the method ultimately destined to succeed; and it is, I believe, the only method which has succeeded down to the present day. Certainly it is the method regularly employed, at Dunsink, at the Cape of Good Hope, and everywhere else where stellar parallax is part of the work.

Between 1830 and 1840 the question was ripe for settlement, and, as frequently happens with a long-matured difficulty, it gave way in three places at once. Bessel, Henderson, and Struve almost simultaneously announced a stellar parallax which could reasonably be accepted. Bessel was a little the earliest, and by far the most accurate. His, indeed, was the result which commanded confidence, and to him the palm must be awarded.

He was largely a self-taught student, having begun life in a counting-house, and having abandoned business for astronomy. But notwithstanding these disadvantages, he became a highly competent mathematician as well as a skilful practical astronomer. He was appointed to superintend the construction of Germany's first great astronomical observatory, that of Königsberg, which, by his system, zeal, and genius, he rapidly made a place of the first importance.

Struve at Dorpat, Bessel at Königsberg, and Henderson at the Cape of Good Hope—all of them at newly-equipped observatories—were severally engaged at the same problem.

But the Russian and German observers had the advantage of the work of one of the most brilliant opticians—I suppose the most brilliant—that has yet appeared: Fraunhofer, of Munich. An orphan lad, apprenticed to a maker of looking-glasses, and subject to hard struggles and privations in early life, he struggled upwards, and ultimately became head of the optical department of a Munich firm of telescope-makers. Here he constructed the famous "Dorpat refractor" for Struve, which is still at work; and designed the "Königsberg heliometer" for Bessel. He also made a long and most skilful research into the solar spectrum, which has immortalized his name. But his health was broken by early trials, and he died at the age of thirty-nine, while planning new and still more important optical achievements.

A heliometer is the most accurate astronomical instrument for relative measurements of position, as a transit circle is the most accurate for absolute determinations. It consists of an equatorial telescope with object-glass cut right across, and each half movable by a sliding movement one past the other, the amount by which the two halves are dislocated being read off by a refined method, and the whole instrument having a multitude of appendages conducive to convenience and accuracy. Its use is to act as a micrometer or measurer of small distances.[28] Each half of the object-glass gives a distinct image, which may be allowed to coincide or may be separated as occasion requires. If it be the components of a double star that are being examined, each component will in general be seen double, so that four images will be seen altogether; but by careful adjustment it will be possible to arrange that one image of each pair shall be superposed on or coincide with each other, in which case only three images are visible; the amount of dislocation of the halves of the object-glass necessary to accomplish this is what is read off. The adjustment is one that can be performed with extreme accuracy, and by performing it again and again with all possible modifications, an extremely accurate determination of the angular distance between the two components is obtained.

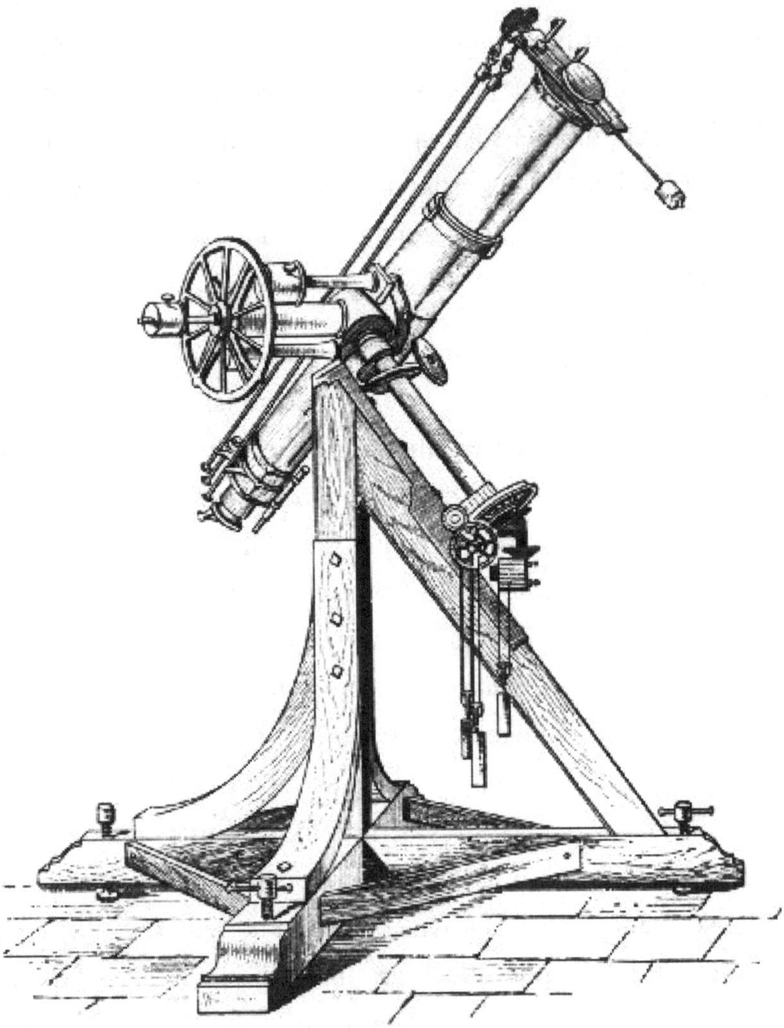

FIG. 92.—Heliometer.

Bessel determined to apply this beautiful instrument to the problem of stellar parallax; and he began by considering carefully the kind of star for which success was most likely. Hitherto the brightest had been most attended to, but Bessel thought that quickness of proper motion would be a still better test of nearness. Not that either criterion is conclusive as to distance, but there was a presumption in favour of either a very bright or an obviously moving star being nearer than a faint or a stationary one; and as the "bright" criterion had already been often applied without result, he decided to try the other. He had already called attention to a record by Piazzi in 1792 of a double star in Cygnus whose proper motion was five seconds of arc every year—a motion which caused this telescopic object, 61 Cygni, to be known as "the flying star." Its motion is not really very perceptible, for it will only have traversed one-third of a lunar diameter in the course of a century; still it was the quickest moving star then known. The position of this interesting double he compared with two other stars which were seen simultaneously in the field of the heliometer, by the method I have described, throughout the whole year 1838; and in the last month of that year he was able to announce with confidence a distinct though very small parallax; substantiating it with a mass of detailed evidence which commanded the assent of astronomers. The amount of it he gave as one-third of a second. We know now that he was very nearly right, though modern research makes it more like half a second.[29]

Soon afterwards, Struve announced a quarter of a second as the parallax of Vega, but that is distinctly too great; and Henderson announced for α Centauri (then thought to be a double) a parallax of one second, which, if correct, would make it quite the nearest of all the stars, but the result is now believed to be about twice too big.

Knowing the distance of 61 Cygni, we can at once tell its real rate of travel—at least, its rate across our line of sight: it is rather over three million miles a day.

Now just consider the smallness of the half second of arc, thus triumphantly though only approximately measured. It is the angle subtended by twenty-six feet at a distance of 2,000 miles. If a telescope planted at New York could be directed to a house in England, and be then turned so as to set its cross-wire first on one end of an ordinary room and then on the other end of the same room, it would have turned through half a second, the angle of greatest stellar parallax. Or, putting it another way. If the star were as near us as New York is, the sun, on the same scale, would be nine paces off. As twenty-six feet is to the distance of New York, so is ninety-two million miles to the distance of the nearest fixed star.

Suppose you could arrange some sort of telegraphic vehicle able to carry you from here to New York in the tenth part of a second—*i.e.* in the time required to drop two inches—such a vehicle would carry you to the moon in twelve seconds, to the sun in an hour and a quarter. Travelling thus continually, in twenty-four hours you would leave the last member of the solar system behind you, and begin your plunge into the depths of space. How long would it be before you encountered another object? A month, should you guess? Twenty years you must journey with that prodigious speed before you reach the nearest star, and then another twenty years before you reach another. At these awful distances from one another the stars are scattered in space, and were they not brilliantly self-luminous and glowing like our sun, they would be hopelessly invisible.

I have spoken of 61 Cygni as a flying star, but there is another which goes still quicker, a faint star, 1830 in Groombridge's Catalogue. Its distance is far greater than that of 61 Cygni, and yet it is seen to move almost as quickly. Its actual speed is about 200 miles a second—greater than the whole visible firmament of fifty million stars can control; and unless the universe is immensely larger than anything we can see with the most powerful telescopes, or unless there are crowds of invisible non-luminous stars mixed up with the others, it can only be a temporary visitor to this frame of things; it is rushing from an infinite distance to an infinite distance; it is passing through our visible universe for the first and only time—it will never return. But so gigantic is the extent of visible space, that even with its amazing speed of 200 miles every second, this star will take two or three million years to get out of sight of our present telescopes, and several thousand years before it gets perceptibly fainter than it is now.

Have we any reason for supposing that the stars we see are all there are? In other words, have we any reason for supposing all celestial objects to be sufficiently luminous to be visible? We have every ground for believing the contrary. Every body in the solar system is dull and dark except the sun, though probably Jupiter is still red-hot. Why may not some of the stars be dark too? The genius of Bessel surmised this, and consistently upheld the doctrine that the astronomy of the future would have to concern itself with dark and invisible bodies; he preached "an astronomy of the invisible." Moreover he predicted the presence of two such dark bodies—one a companion of Sirius, the other of Procyon. He noticed certain irregularities in the motions of these stars which he asserted must be caused by their revolving round other bodies in a period of half a century. He announced in 1844 that both Sirius and Procyon were double stars, but that their companions, though large, were dark, and therefore invisible.

No one accepted this view, till Peters, in America, found in 1851 that the hypothesis accurately explained the anomalous motion of Sirius, and, in fact, indicated an exact place where the companion ought to be. The obscure companion of Sirius became now a recognized celestial object, although it had never been seen, and it was held to revolve round Sirius in fifty years, and to be about half as big.

In 1862, the firm of Alvan Clark and Sons, of New York, were completing a magnificent 18-inch refractor, and the younger Clark was trying it on Sirius, when he said: "Why, father, the star has a companion!" The elder Clark also looked, and sure enough there was a faint companion due east of the bright star, and in just the position required by theory. Not that the Clarks knew anything about the theory. They were keen-sighted and most skilful instrument-makers, and they made the discovery by accident. After it had once been seen, it was found that several of the large telescopes of the world were able to show it. It is half as big, but it only gives 1/10000th part of the light that Sirius gives. No doubt it shines partly with a borrowed light and partly with a dull heat of its own. It is a real planet, but as yet too hot to live on. It will cool down in time, as our earth has cooled and as Jupiter is cooling, and no doubt become habitable enough. It does revolve round Sirius in a period of 49·4 years—almost exactly what Bessel assigned to it.

But Bessel also assigned a dark companion to Procyon. It and its luminous neighbour are considered to revolve round each other in a period of forty years, and astronomers feel perfectly assured of its existence, though at present it has not been seen by man.

LECTURE XV

THE DISCOVERY OF NEPTUNE

WE approach to-night perhaps the greatest, certainly the most conspicuous, triumphs of the theory of gravitation. The explanation by Newton of the observed facts of the motion of the moon, the way he accounted for precession and nutation and for the tides, the way in which Laplace explained every detail of the planetary motions—these achievements may seem to the professional astronomer equally, if not more, striking and wonderful; but of the facts to be explained in these cases the general public are necessarily more or less ignorant, and so no beauty or thoroughness of treatment appeals to them, nor can excite their imaginations. But to predict in the solitude of the study, with no weapons other than pen, ink, and paper, an unknown and enormously distant world, to calculate its orbit when as yet it had never been seen, and to be able to say to a practical astronomer, "Point your telescope in such a direction at such a time, and you will see a new planet hitherto unknown to man"—this must always appeal to the imagination with dramatic intensity, and must awaken some interest in almost the dullest.

Prediction is no novelty in science; and in astronomy least of all is it a novelty. Thousands of years ago, Thales, and others whose very names we have forgotten, could predict eclipses with some certainty, though with only rough accuracy. And many other phenomena were capable of prediction by accumulated experience. We have seen, for instance (coming to later times), how a gap between Mars and Jupiter caused a missing planet to be suspected and looked for, and to be found in a hundred pieces. We have seen, also, how the abnormal proper-motion of Sirius suggested to Bessel the existence of an unseen companion. And these last instances seem to approach very near the same class of prediction as that of the discovery of Neptune. Wherein, then, lies the difference? How comes it that some classes of prediction—such as that if you put your finger in fire it will get burnt—are childishly easy and commonplace, while others excite in the keenest intellects the highest feelings of admiration? Mainly, the difference lies, first, in the grounds on which the prediction is based; second, on the difficulty of the investigation whereby it is accomplished; third, in the completeness and the accuracy with which it can be verified. In all these points, the discovery of Neptune stands out pre-eminently among the verified predictions of science, and the circumstances surrounding it are of singular interest.

In 1781, Sir William Herschel discovered the planet Uranus. Now you know that three distinct observations suffice to determine the orbit of a planet completely, and that it is well to have the three observations as far apart as possible so as to minimize the effects of minute but necessary errors of observation. (See p. 298.) Directly Uranus was found, therefore, old records of stellar observations were ransacked, with the object of discovering whether it had ever been unwittingly seen before. If seen, it had been thought of course to be a star (for it shines like a star of the sixth magnitude, and can therefore be just seen without a telescope if one knows precisely where to look for it, and if one has good sight), but if it had been seen and catalogued as a star it would have moved from its place, and the catalogue would by that entry be wrong. The thing to detect, therefore, was errors in the catalogues: to examine all entries, and see if the stars entered actually existed, or were any of them missing. If a wrong entry were discovered, it might of course have been due to some clerical error, though that is hardly probable considering the care taken over these things, or it might have been some tailless comet or other, or it might have been the newly found planet.

So the next thing was to calculate backwards, and see if by any possibility the planet could have been in that place at that time. Examined in this way the tabulated observations of Flamsteed showed that he had unwittingly observed Uranus five distinct times, the first time in 1690, nearly a century before Herschel discovered its true nature. But more remarkable still, Le Monnier, of Paris, had observed it eight times in one month, cataloguing it each time as a different star. If only he had reduced and compared his observations, he would have anticipated Herschel by twelve years. As it was, he missed it altogether. It was seen once by Bradley also. Altogether it had been seen twenty times.

These old observations of Flamsteed and those of Le Monnier, combined with those made after Herschel's discovery, were very useful in determining an exact orbit for the new planet, and its motion was considered thoroughly known. It was not an *exact* ellipse, of course: none of the planets describe *exact* ellipses—each perturbs all the rest, and these small perturbations must be taken into account, those of Jupiter and Saturn being by far the most important.

For a time Uranus seemed to travel regularly and as expected, in the orbit which had been calculated for it; but early in the present century it began to be slightly refractory, and by 1820 its actual place showed quite a distinct discrepancy from its position as calculated with the aid of the old observations. It was at first thought that this discrepancy must be due to inaccuracies in the older observations, and they were accordingly rejected, and tables prepared for the planet based on the newer and more accurate observations only. But by 1830 it became apparent that it would not accurately obey even these. The error amounted to some 20". By 1840 it was as much as 90', or a minute and a half. This discrepancy is quite distinct, but still it is very small, and had two objects been in the heavens at once, the actual Uranus and the theoretical Uranus, no unaided eye could possibly have distinguished them or detected that they were other than a single star.

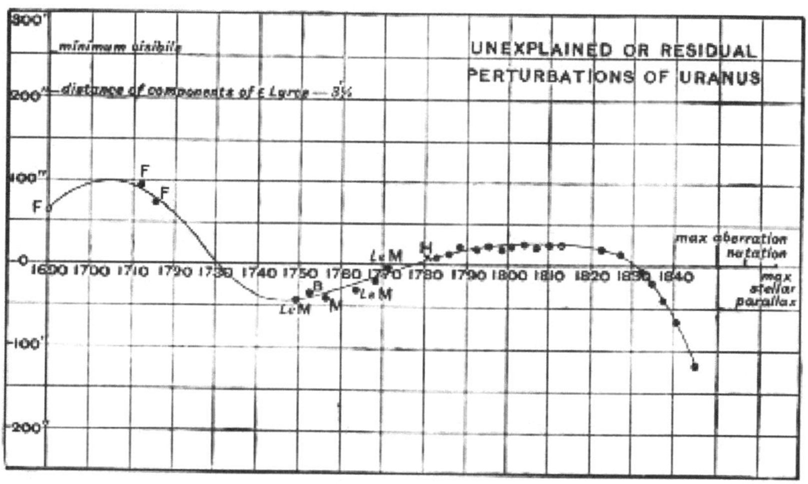

FIG. 93.—Perturbations of Uranus.

The chance observations by Flamsteed, by Le Monnier, and others, are plotted in this diagram, as well as the modern determinations made after Herschel had discovered the nature of the planet. The decades are laid off horizontally. Vertical distance represents the difference between observed and subsequently calculated longitudes—in other words, the principal perturbations caused by Neptune. To show the scale, a number of standard things are represented too by lengths measured upwards from the line of time, viz: the smallest quantity perceptible to the naked eye,—the maximum angle of aberration, of nutation, and of stellar parallax; though this last is too small to be properly indicated. The perturbations are much bigger than these; but compared with what can be seen without a telescope they are small—the distance between the component pairs of ε Lyræ (210") (see fig. 86, page 288), which a few keen-eyed persons can see as a simple double star, being about twice the greatest perturbation.

The diagram shows all the irregularities plotted in the light of our present knowledge; and, to compare with their amounts, a few standard things are placed on the same scale, such as the smallest interval capable of being detected with the unaided eye, the distance of the component stars in ε Lyræ, the constants of aberration, of nutation, and of stellar parallax.

The errors of Uranus therefore, though small, were enormously greater than things which had certainly been observed; there was an unmistakable discrepancy between theory and observation. Some cause was evidently at work on this distant planet, causing it to disagree with its motion as calculated according to the law of gravitation. Some thought that the exact law of gravitation did not apply to so distant a body. Others surmised the presence of some foreign and unknown body, some comet, or some still more distant planet perhaps, whose gravitative attraction for Uranus was the cause of the whole difficulty—some perturbations, in fact, which had not been taken into account because of our ignorance of the existence of the body which caused them.

But though such an idea was mentioned among astronomers, it was not regarded with any special favour, and was considered merely as one among a number of hypotheses which could be suggested as fairly probable.

It is perfectly right not to attach much importance to unelaborated guesses. Not until the consequences of an hypothesis have been laboriously worked out—not until it can be shown capable of producing the effect quantitatively as well as qualitatively—does its statement rise above the level of a guess, and attain the dignity of a theory. A later stage still occurs when the theory has been actually and completely verified by agreement with observation.

Now the errors in the motion of Uranus, *i.e.* the discrepancy between its observed and calculated longitudes—all known disturbing causes, such as Jupiter and Saturn, being allowed for—are as follows (as quoted by Dr. Haughton) in seconds of arc:—

ANCIENT OBSERVATIONS (casually made, as of a star).

Flamsteed	1690	+61·2
"	1712	+92·7
"	1715	+73·8
Le Monnier	1750	-47·6
Bradley	1753	-39·5
Mayer	1756	-45·7
Le Monnier	1764	-34·9
"	1769	-19·3
"	1771	-2·3

Modern Observations.	
1780	+3·46
1783	+8·45
1786	+12·36
1789	+19·02
1801	+22·21
1810	+23·16
1822	+20·97
1825	+18·16
1828	+10·82
1831	-3·98
1834	-20·80
1837	-42·66
1840	-66·64

These are the numbers plotted in the above diagram (Fig. 92), where H marks the discovery of the planet and the beginning of its regular observation.

Something was evidently the matter with the planet. If the law of gravitation held exactly at so great a distance from the sun, there must be some perturbing force acting on it besides all those known ones which had been fully taken into account. Could it be an outer planet? The question occurred to several, and one or two tried if they could solve the problem, but were soon stopped by the tremendous difficulties of calculation.

The ordinary problem of perturbation is difficult enough: Given a disturbing planet in such and such a position, to find the perturbations it produces. This problem it was that Laplace worked out in the *Mécanique Céleste.*

But the inverse problem: Given the perturbations, to find the planet which causes them—such a problem had never yet been attacked, and by only a few had its possibility been conceived. Bessel made preparations for trying what he could do at it in 1840, but he was prevented by fatal illness.

In 1841 the difficulties of the problem presented by these residual perturbations of Uranus excited the imagination of a young student, an undergraduate of St. John's College, Cambridge—John Couch Adams by name—and he determined to have a try at it as soon as he was through his Tripos. In January, 1843, he graduated as Senior Wrangler, and shortly afterwards he set to work. In less than two years he reached a definite conclusion; and in October, 1845, he wrote to the Astronomer-Royal, at Greenwich, Professor Airy, saying that the perturbations of Uranus would be explained by assuming the existence of an outer planet, which he reckoned was now situated in a specified latitude and longitude.

We know now that had the Astronomer-Royal put sufficient faith in this result to point his big telescope to the spot indicated and commence sweeping for a planet, he would have detected it within 1¾° of the place assigned to it by Mr. Adams. But any one in the position of the Astronomer-Royal knows that almost every post brings an absurd letter from some ambitious correspondent or other, some of them having just discovered perpetual motion, or squared the circle, or proved the earth flat, or discovered the constitution of the moon, or of ether, or of electricity; and out of this mass of rubbish it requires great skill and patience to detect such gems of value as there may be.

Now this letter of Mr. Adams's was indeed a jewel of the first water, and no doubt bore on its face a very different appearance from the chaff of which I have spoken; but still Mr. Adams was an unknown man: he had graduated as Senior Wrangler it is true, but somebody must graduate as Senior Wrangler every year, and every year by no means produces a first-rate mathematician. Those behind the scenes, as Professor Airy of course was, having been a Senior Wrangler himself, knew perfectly well that the labelling of a young man on taking his degree is much more worthless as a testimony to his genius and ability than the general public are apt to suppose.

Was it likely that a young and unknown man should have successfully solved so extremely difficult a problem? It was altogether unlikely. Still, he would test him: he would ask for further explanations concerning some of the perturbations which he himself had specially noticed, and see if Mr. Adams could explain these also by his hypothesis. If he could, there might be something in his theory. If he failed—well, there was an end of it. The questions were not difficult. They concerned the error of the radius vector. Mr. Adams could have answered them with perfect ease; but sad to say, though a brilliant mathematician, he was not a man of business. He did not answer Professor Airy's letter.

It may to many seem a pity that the Greenwich Equatoreal was not pointed to the place, just to see whether any foreign object did happen to be in that neighbourhood; but it is no light matter to derange the work of an Observatory, and alter the work mapped out for the staff into a sudden sweep for a new planet, on the strength of a mathematical investigation just received by post. If observatories were conducted on these unsystematic and spasmodic principles, they would not be the calm, accurate, satisfactory places they are.

Of course, if any one could have known that a new planet was to be had for the looking, *any* course would have been justified; but no one could know this. I do not suppose that Mr. Adams himself could feel all that confidence in his attempted prediction. So there the matter dropped. Mr. Adams's communication was pigeon-holed, and remained in seclusion for eight or nine months.

Meanwhile, and quite independently, something of the same sort was going on in France. A brilliant young mathematician, born in Normandy in 1811, had accepted the post of Astronomical Professor at the École Polytechnique, then recently founded by Napoleon. His first published papers directed attention to his wonderful powers; and the official head of astronomy in France, the famous Arago, suggested to him the unexplained perturbations of Uranus as a worthy object for his fresh and well-armed vigour.

At once he set to work in a thorough and systematic way. He first considered whether the discrepancies could be due to errors in the tables or errors in the old observations. He discussed them with minute care, and came to the conclusion that they were not thus to be explained away. This part of the work he published in November, 1845.

He then set to work to consider the perturbations produced by Jupiter and Saturn, to see if they had been with perfect accuracy allowed for, or whether some minute improvements could be made sufficient to destroy the irregularities. He introduced several fresh terms into these perturbations, but none of them of sufficient magnitude to do more than slightly lessen the unexplained perturbations.

He next examined the various hypotheses that had been suggested to account for them:—Was it a failure in the law of gravitation? Was it due to the presence of a resisting medium? Was it due to some unseen but large satellite? Or was it due to a collision with some comet?

All these he examined and dismissed for various reasons one after the other. It was due to some steady continuous cause—for instance, some unknown planet. Could this planet be inside the orbit of Uranus? No, for then it would perturb Saturn and Jupiter also, and they were not perturbed by it. It must, therefore, be some planet outside the orbit of Uranus, and in all probability, according to Bode's empirical law, at nearly double the distance from the sun that Uranus is. Lastly he proceeded to examine where this planet was, and what its orbit must be to produce the observed disturbances.

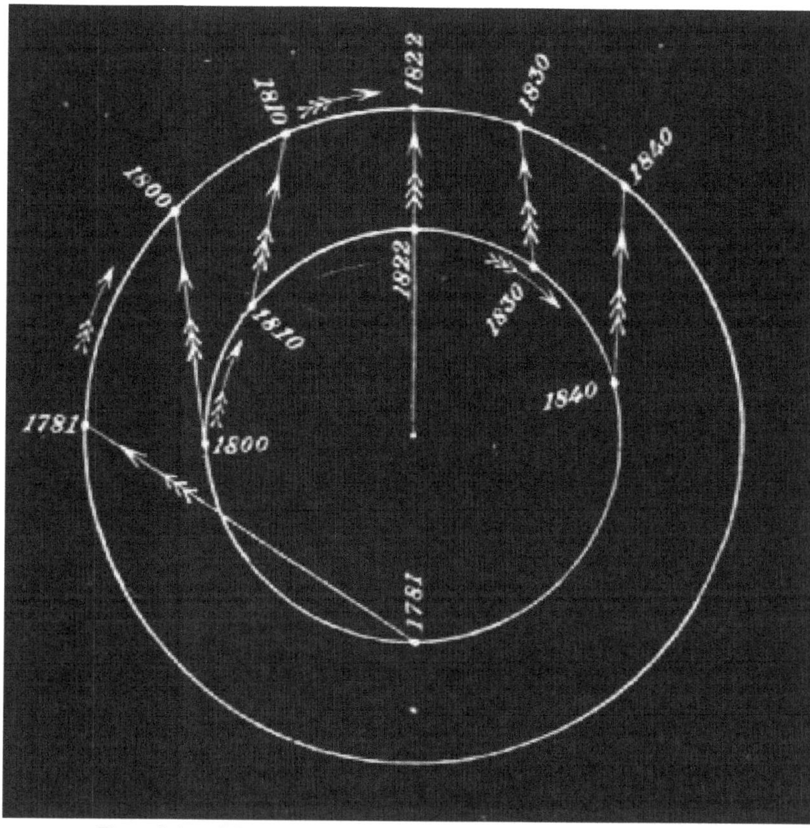

FIG. 94.—Uranus's and Neptune's relative positions.

The above diagram, drawn to scale by Dr. Haughton, shows the paths of Uranus and Neptune, and their positions from 1781 to 1840, and illustrates the *direction* of their mutual perturbing force. In 1822 the planets were in conjunction, and the force would then perturb the radius vector (or distance from the sun), but not the longitude (or place in orbit). Before that date Uranus had been hurried along, and after that date it had been retarded, by the pull of Neptune, and thus the observed discrepancies from its computed place were produced. The problem was first to disentangle the outstanding perturbations from those which would be caused by Jupiter and Saturn and all other known causes, and then to assign the place of an outer planet able to produce precisely those perturbations in Uranus.

Not without failures and disheartening complications was this part of the process completed. This was, after all, the real tug of war. So many unknown quantities: its mass, its distance, its excentricity, the obliquity of its orbit, its position at any time—nothing known, in fact, about the planet except the microscopic disturbance it caused in Uranus, some thousand million miles away from it.

Without going into further detail, suffice it to say that in June, 1846, he published his last paper, and in it announced to the world his theoretical position for the planet.

Professor Airy received a copy of this paper before the end of the month, and was astonished to find that Leverrier's theoretical place for the planet was within 1° of the place Mr. Adams had assigned to it eight months before. So striking a coincidence seemed sufficient to justify a Herschelian "sweep" for a week or two.

But a sweep for so distant a planet would be no easy matter. When seen in a large telescope it would still only look like a star, and it would require considerable labour and watching to sift it out from the other stars surrounding it. We know that Uranus had been seen twenty times, and thought to be a star, before its true nature was by Herschel discovered; and Uranus is only about half as far away as Neptune is.

Neither in Paris nor yet at Greenwich was any optical search undertaken; but Professor Airy wrote to ask M. Leverrier the same old question as he had fruitlessly put to Mr. Adams: Did the new theory explain the errors of the radius vector or not? The reply of Leverrier was both prompt and satisfactory—these errors were explained, as well as all the others. The existence of the object was then for the first time officially believed in.

The British Association met that year at Southampton, and Sir John Herschel was one of its Sectional Presidents. In his inaugural address, on September 10th, 1846, he called attention to the researches of Leverrier and Adams in these memorable words:—

"The past year has given to us the new [minor] planet Astræa; it has done more—it has given us the probable prospect of another. We see it as Columbus saw America from the shores of Spain. Its movements have been felt trembling along the far-reaching line of our analysis with a certainty hardly inferior to ocular demonstration."

It was about time to begin to look for it. So the Astronomer-Royal thought on reading Leverrier's paper. But as the national telescope at Greenwich was otherwise occupied, he wrote to Professor Challis, at Cambridge, to know if he would permit a search to be made for it with the Northumberland Equatoreal, the large telescope of Cambridge University, presented to it by one of the Dukes of Northumberland.

Professor Challis said he would conduct the search himself; and shortly commenced a leisurely and dignified series of sweeps round about the place assigned by theory, cataloguing all the stars which he observed, intending afterwards to sort out his observations, compare one with another, and find out whether any one star had changed its position; because if it had it must be the planet. He thus, without giving an excessive time to the business, accumulated a host of observations, which he intended afterwards to reduce and sift at his leisure.

The wretched man thus actually saw the planet twice—on August 4th and August 12th, 1846—without knowing it. If only he had had a map of the heavens containing telescopic stars down to the tenth magnitude, and if he had compared his observations with this map as they were made, the process would have been easy, and the discovery quick. But he had no such map. Nevertheless one was in existence: it had just been completed in that country of enlightened method and industry—Germany. Dr. Bremiker had not, indeed, completed his great work—a chart of the whole zodiac down to stars of the tenth magnitude—but portions of it were completed, and the special region where the new planet was expected happened to be among the portions already just done. But in England this was not known.

Meanwhile, Mr. Adams wrote to the Astronomer-Royal several additional communications, making improvements in his theory, and giving what he considered nearer and nearer approximations for the place of the planet. He also now answered quite satisfactorily, but too late, the question about the radius vector sent to him months before.

Let us return to Leverrier. This great man was likewise engaged in improving his theory and in considering how best the optical search could be conducted. Actuated, probably, by the knowledge that in such matters as cataloguing and mapping Germany was then, as now, far ahead of all the other nations of the world, he wrote in September (the same September as Sir John Herschel delivered his eloquent address at Southampton) to Berlin. Leverrier wrote, I say, to Dr. Galle, head of the Observatory at Berlin, saying to him, clearly and decidedly, that the new planet was now in or close to such and such a position, and that if he would point his telescope to that part of the heavens he would see it; and, moreover, that he would be able to tell it from a star by its having a sensible magnitude, or disk, instead of being a mere point.

Galle got the letter on the 23rd of September, 1846. That same evening he did point his telescope to the place Leverrier told him, and he saw the planet that very night. He recognized it first by its appearance. To his practised eye it did seem to have a small disk, and not quite the same aspect as an ordinary star. He then consulted Bremiker's great star chart, the part just engraved and finished, and sure enough on that chart there was no such star there. Undoubtedly it was the planet.

The news flashed over Europe at the maximum speed with which news could travel at that date (which was not very fast); and by the 1st of October Professor Challis and Mr. Adams heard it at Cambridge, and had the pleasure of knowing that they were forestalled, and that England was out of the race.

It was an unconscious race to all concerned, however. Those in France knew nothing of the search going on in England. Mr. Adams's papers had never been published; and very annoyed the French were when a claim was set up on his behalf to a share in this magnificent discovery. Controversies and recriminations, excuses and justifications, followed; but the discussion has now settled down. All the world honours the bright genius and mathematical skill of Mr. Adams, and recognizes that he first solved the problem by calculation. All the world, too, perceives clearly the no less eminent mathematical talents of M. Leverrier, but it recognizes in him something more than the mere mathematician—the man of energy, decision, and character.

LECTURE XVI

COMETS AND METEORS

We have now considered the solar system in several aspects, and we have passed in review something of what is known about the stars. We have seen how each star is itself, in all probability, the centre of another and distinct solar system, the constituents of which are too dark and far off to be visible to us; nothing visible here but the central sun alone, and that only as a twinkling speck.

But between our solar system and these other suns—between each of these suns and all the rest—there exist vast empty spaces, apparently devoid of matter.

We have now to ask, Are these spaces really empty? Is there really nothing in space but the nebulæ, the suns, their planets, and their satellites? Are all the bodies in space of this gigantic size? May there not be an infinitude of small bodies as well?

The answer to this question is in the affirmative. There appears to be no special size suited to the vastness of space; we find, as a matter of fact, bodies of all manner of sizes, ranging by gradations from the most tremendous suns, like Sirius, down through ordinary suns to smaller ones, then to planets of all sizes, satellites still smaller, then the asteroids, till we come to the smallest satellite of Mars, only about ten miles in diameter, and weighing only some billion tons—the smallest of the regular bodies belonging to the solar system known.

But, besides all these, there are found to occur other masses, not much bigger and some probably smaller, and these we call comets when we see them. Below these, again, we find masses varying from a few tons in weight down to only a few pounds or ounces, and these when we see them, which is not often, we call meteors or shooting-stars; and to the size of these meteorites there would appear to be no limit: some may be literal grains of dust. There seems to be a regular gradation of size, therefore, ranging from Sirius to dust; and apparently we must regard all space as full of these cosmic particles—stray fragments, as it were, perhaps of some older world, perhaps going to help to form a new one some day. As Kepler said, there are more "comets" in the sky than fish in the sea. Not that they are at all crowded together, else they would make a cosmic haze. The transparency of space shows that there must be an enormous proportion of clear space between each, and they are probably much more concentrated near one of the big bodies than they are in interstellar space.[30] Even during the furious hail of meteors in November 1866 it was estimated that their average distance apart in the thickest of the shower was 35 miles.

Consider the nature of a meteor or shooting-star. We ordinarily see them as a mere streak of light; sometimes they leave a luminous tail behind them; occasionally they appear as an actual fire-ball, accompanied by an explosion; sometimes, but very seldom, they are seen to drop, and may subsequently be dug up as a lump of iron or rock, showing signs of rough treatment by excoriation and heat. These last are the meteorites, or siderites, or aërolites, or bolides, of our museums. They are popularly spoken of as thunderbolts, though they have nothing whatever to do with atmospheric electricity.

FIG. 95.—Meteorite.

They appear to be travelling rocky or metallic fragments which in their journey through space are caught in the earth's atmosphere and instantaneously ignited by the friction. Far away in the depths of space one of these bodies felt the attracting power of the sun, and began moving towards him. As it approached, its speed grew gradually quicker and quicker continually, until by the time it has approached to within the distance of the earth, it whizzes past with the velocity of twenty-six miles a second. The earth is moving on its own account nineteen miles every second. If the two bodies happened to be moving in opposite directions, the combined speed would be terrific; and the faintest trace of atmosphere, miles above the earth's surface, would exert a furious grinding action on the stone. A stream of particles would be torn off; if of iron, they would burn like a shower of filings from a firework, thus forming a trail; and the mass itself would be dissipated, shattered to fragments in an instant.

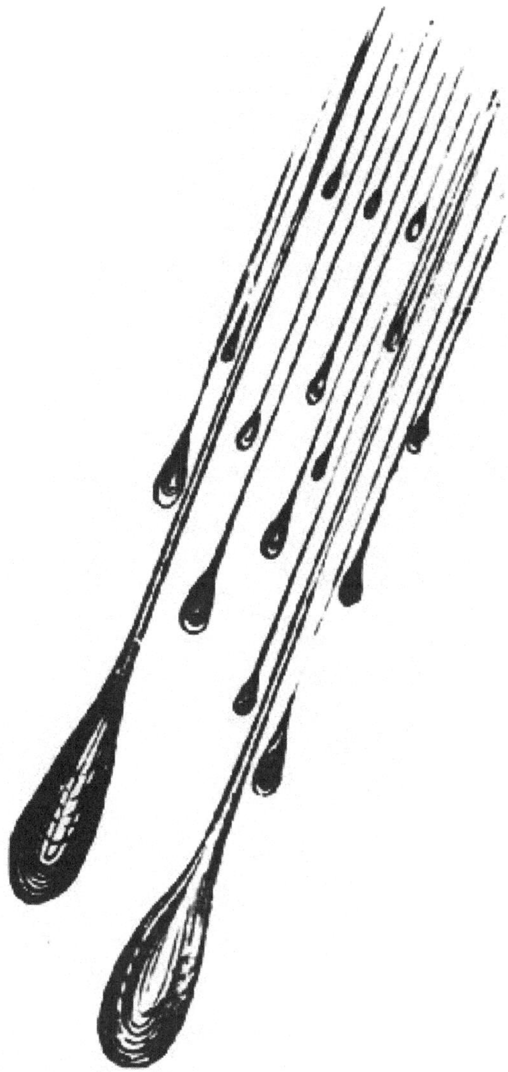

FIG. 96.—Meteor stream crossing field of telescope.

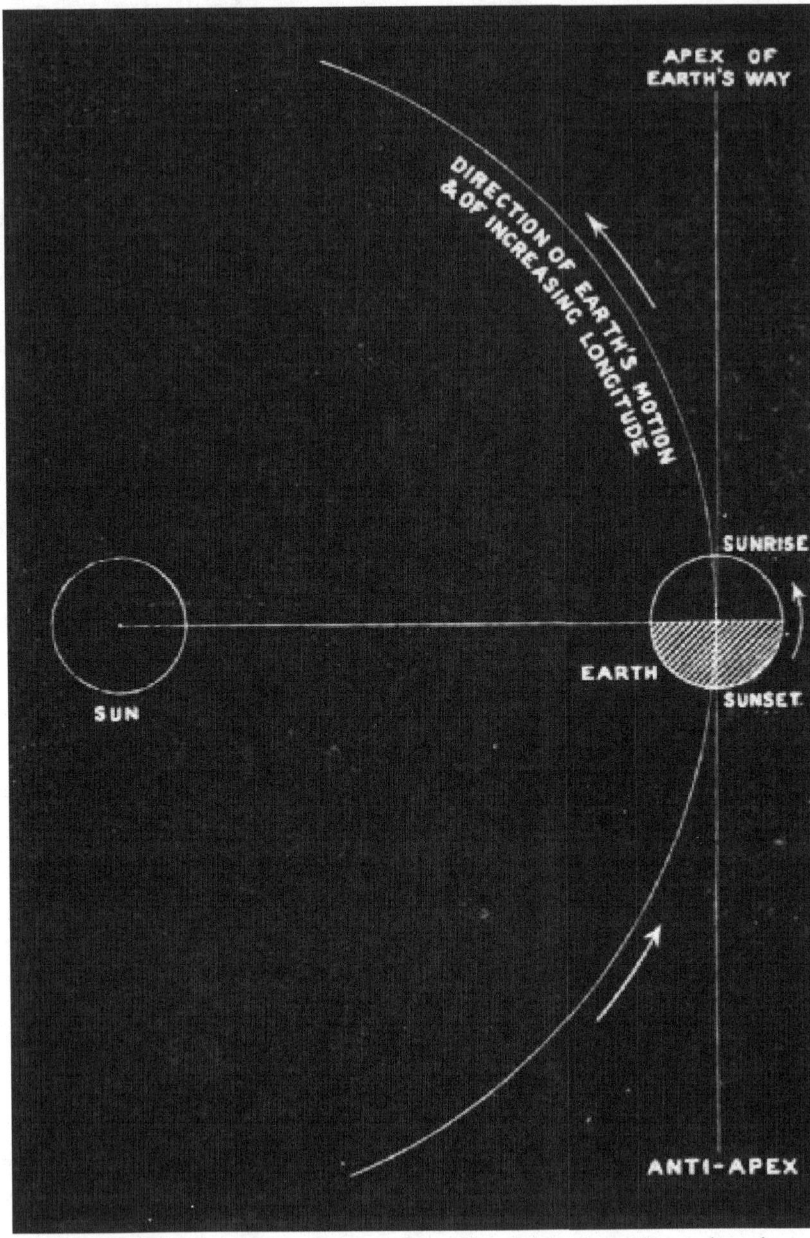

FIG. 97.—Diagram of direction of earth's orbital motion, showing that after midnight, *i.e.* between midnight and noon, more asteroids are

likely to be swept up by any locality than between noon and midnight. [From Sir R.S. Ball.]

Even if the earth were moving laterally, the same thing would occur. But if earth and stone happened to be moving in the same direction, there would be only the differential velocity of seven miles a second; and though this is in all conscience great enough, yet there might be a chance for a residue of the nucleus to escape entire destruction, though it would be scraped, heated, and superficially molten by the friction; but so much of its speed would be rubbed out of it, that on striking the earth it might bury itself only a few feet or yards in the soil, so that it could be dug out. The number of those which thus reach the earth is comparatively infinitesimal. Nearly all get ground up and dissipated by the atmosphere; and fortunate it is for us that they are so. This bombardment of the exposed face of the moon must be something terrible.[31]

Thus, then, every shooting-star we see, and all the myriads that we do not and cannot see because they occur in the day-time, all these bright flashes or streaks, represent the death and burial of one of these flying stones. It had been careering on its own account through space for untold ages, till it meets a planet. It cannot strike the actual body of the planet—the atmosphere is a sufficient screen; the tremendous friction reduces it to dust in an instant, and this dust then quietly and leisurely settles down on to the surface.

Evidence of the settlement of meteoric dust is not easy to obtain in such a place as England, where the dust which accumulates is seldom of a celestial character; but on the snow-fields of Greenland or the Himalayas dust can be found; and by a Committee of the British Association distinct evidence of molten globules of iron and other materials appropriate to aërolites has been obtained, by the simple process of collecting, melting, and filtering long exposed snow. Volcanic ash may be mingled with it, but under the microscope the volcanic and the meteoric constituents have each a distinctive character.

The quantity of meteoric material which reaches the earth as dust must be immensely in excess of the minute quantity which arrives in the form of lumps. Hundreds or thousands of tons per annum must be received; and the accretion must, one would think, in the course of ages be able to exert some influence on the period of the earth's rotation—the length of the day. It is too small, however, to have been yet certainly detected. Possibly, it is altogether negligible.

It has been suggested that those stones which actually fall are not the true cosmic wanderers, but are merely fragments of our own earth, cast up by powerful volcanoes long ago when the igneous power of the earth was more vigorous than now—cast up with a speed of close upon seven miles a second; and now in these quiet times gradually being swept up by the earth, and so returning whence they came.

I confess I am unable to draw a clear distinction between one set and the other. Some falling stars may have had an origin of this sort, but certainly others have not; and it would seem very unlikely that one set only should fall bodily upon the earth, while the others should always be rubbed to powder. Still, it is a possibility to be borne in mind.

We have spoken of these cosmic visitors as wandering masses of stone or iron; but we should be wrong if we associated with the term "wandering" any ideas of lawlessness and irregularity of path. These small lumps of matter are as obedient to the law of gravity as any large ones can be. They must all, therefore, have definite orbits, and these orbits will have reference to the main attracting power of our system—they will, in fact, be nearly all careering round the sun.

Each planet may, in truth, have a certain following of its own. Within the limited sphere of the earth's predominant attraction, for instance, extending some way beyond the moon, we may have a number of satellites that we never see, all revolving regularly in elliptic orbits round the earth. But, comparatively speaking, these satellite meteorites are few. The great bulk of them will be of a planetary character—they will be attendant upon the sun.

It may seem strange that such minute bodies should have regular orbits and obey Kepler's laws, but they must. All three laws must be as rigorously obeyed by them as by the planets themselves. There is nothing in the smallness of a particle to excuse it from implicit obedience to law. The only consequence of their smallness is their inability to perturb others. They cannot appreciably perturb either the planets they approach or each other. The attracting power of a lump one million tons in weight is very minute. A pound, on the surface of such a body of the same density as the earth, would be only pulled to it with a force equal to that with which the earth pulls a grain. So the perturbing power of such a mass on distant bodies is imperceptible. It is a good thing it is so: accurate astronomy would be impossible if we had to take into account the perturbations caused by a crowd of invisible bodies. Astronomy would then approach in complexity some of the problems of physics.

But though we may be convinced from the facts of gravitation that these meteoric stones, and all other bodies flying through space near our solar system, must be constrained by the sun to obey Kepler's laws, and fly round it in some regular elliptic or hyperbolic orbit, what chance have we of determining that orbit? At first sight, a very poor chance, for we never see them except for the instant when they splash into our atmosphere; and for them that instant is instant death. It is unlikely that any escape that ordeal, and even if they do, their career and orbit are effectually changed. Henceforward they must become attendants on the earth. They may drop on to its surface, or they may duck out of our atmosphere again, and revolve round us unseen in the clear space between earth and moon.

Nevertheless, although the problem of determining the original orbit of any given set of shooting-stars before it struck us would seem nearly insoluble, it has been solved, and solved with some approach to accuracy; being done by the help of observations of certain other bodies. The bodies by whose help this difficult problem has been attacked and resolved are comets. What are comets?

I must tell you that the scientific world is not entirely and completely decided on the structure of comets. There are many floating ideas on the subject, and some certain knowledge. But the subject is still, in many respects, an open one, and the ideas I propose to advocate you will accept for no more than they are worth, viz. as worthy to be compared with other and different views.

Up to the time of Newton, the nature of comets was entirely unknown. They were regarded with superstitious awe as fiery portents, and were supposed to be connected with the death of some king, or with some national catastrophe.

Even so late as the first edition of the *Principia* the problem of comets was unsolved, and their theory is not given; but between the first and the second editions a large comet appeared, in 1680, and Newton speculated on its appearance and behaviour. It rushed down very close to the sun, spun half round him very quickly, and then receded from him again. If it were a material substance, to which the law of gravitation applied, it must be moving in a conic section with the sun in one focus, and its radius vector must sweep out equal areas in equal times. Examining the record of its positions made at observatories, he found its observed path quite accordant with theory; and the motion of comets was from that time understood. Up to that time no one had attempted to calculate an orbit for a comet. They had been thought irregular and lawless bodies. Now they were recognized as perfectly obedient to the law of gravitation, and revolving round the sun like everything else—as members, in fact, of our solar system, though not necessarily permanent members.

But the orbit of a comet is very different from a planetary one. The excentricity of its orbit is enormous—in other words, it is either a very elongated ellipse or a parabola. The comet of 1680, Newton found to move in an orbit so nearly a parabola that the time of describing it must be reckoned in hundreds of years at the least. It is now thought possible that it may not be quite a parabola, but an ellipse so elongated that it will not return till 2255. Until that date arrives, however, uncertainty will prevail as to whether it is a periodic comet, or one of those that only visit our system once. If it be periodic, as suspected, it is the same as appeared when Julius Cæsar was killed, and which likewise appeared in the years 531 and 1106 A.D. Should it appear in 2255, our posterity will probably regard it as a memorial of Newton.

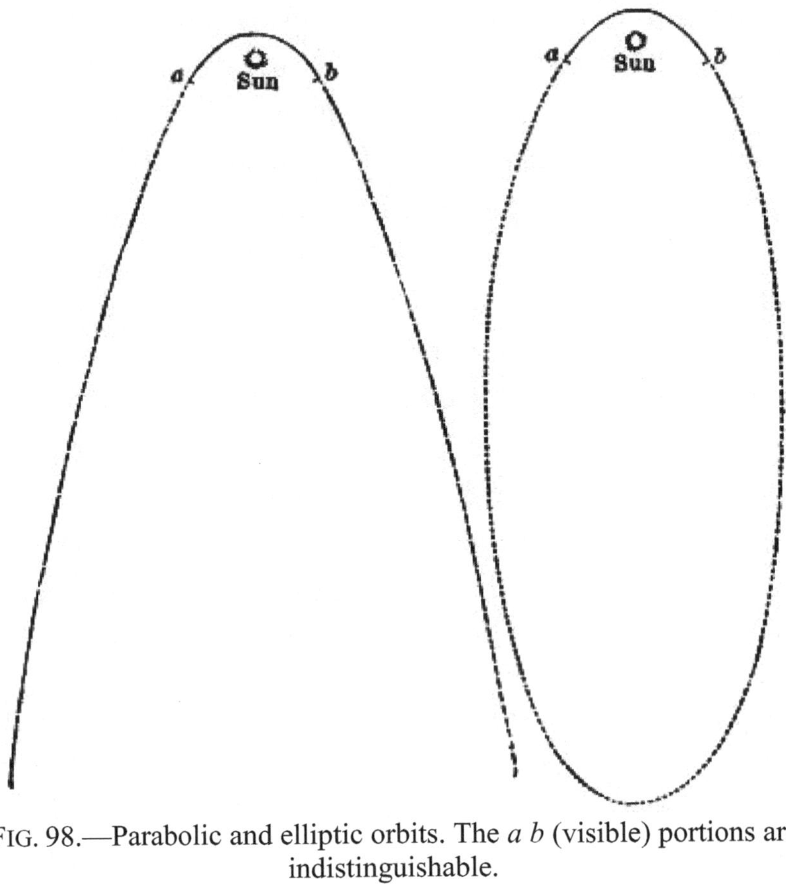

FIG. 98.—Parabolic and elliptic orbits. The *a b* (visible) portions are indistinguishable.

The next comet discussed in the light of the theory of gravitation was the famous one of Halley. You know something of the history of this. Its period is 75½ years. Halley saw it in 1682, and predicted its return in 1758 or 1759—the first cometary prediction. Clairaut calculated its return right within a month (p. 219). It has been back once more, in 1835; and this time its date was correctly predicted within three days, because Uranus was now known. It was away at its furthest point in 1873. It will be back again in 1911.

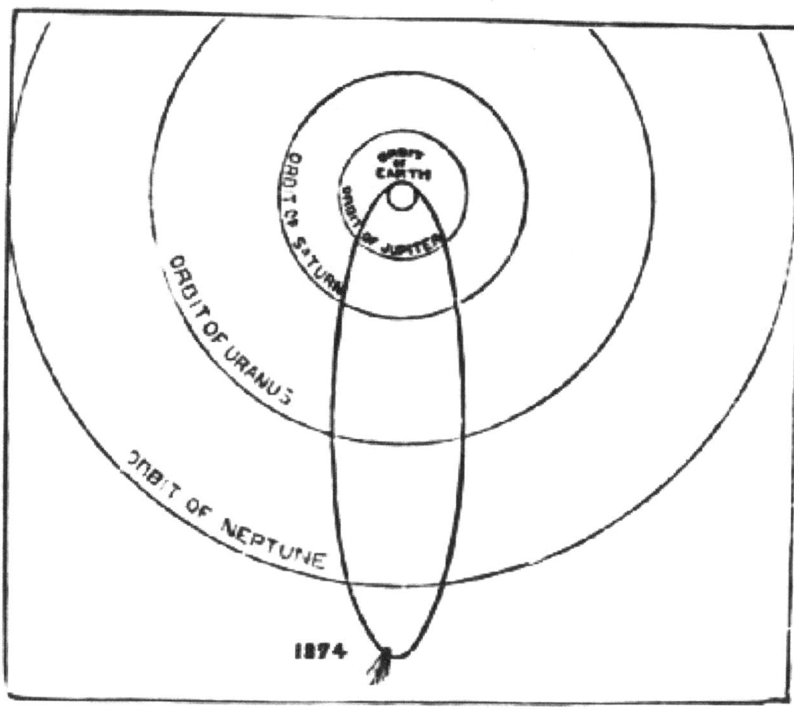

FIG. 99.—Orbit of Halley's comet.

Coming to recent times, we have the great comets of 1843 and of 1858, the history of neither being known. Quite possibly they arrived then for the first time. Possibly the second will appear again in 3808. But besides these great comets, there are a multitude of telescopic ones, which do not show these striking features, and have no gigantic tail. Some have no tail at all, others have at best a few insignificant streamers, and others show a faint haze looking like a microscopic nebula.

All these comets are of considerable extent—some millions of miles thick usually, and yet stars are clearly visible through them. Hence they must be matter of very small density; their tails can be nothing more dense than a filmy mist, but their nucleus must be something more solid and substantial.

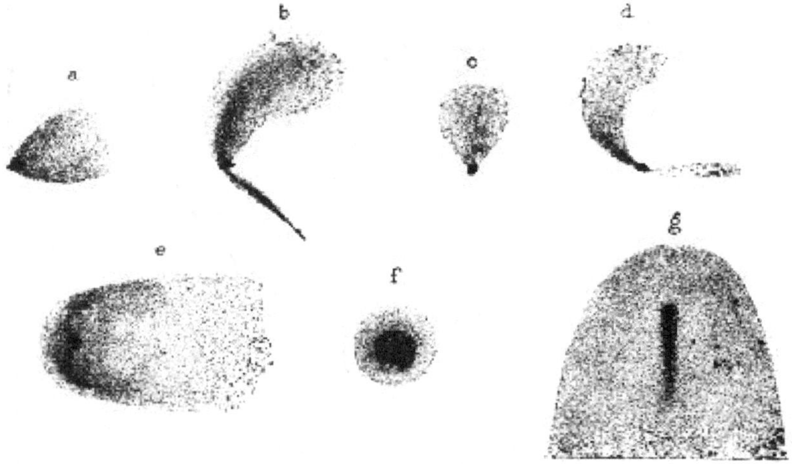

FIG. 100.—Various appearances of Halley's comet when last seen.

I have said that comets arrive from the depths of space, rush towards and round the sun, whizzing past the earth with a speed of twenty-six miles a second, on round the sun with a far greater velocity than that, and then rush off again. Now, all the time they are away from the sun they are invisible. It is only as they get near him that they begin to expand and throw off tails and other appendages. The sun's heat is evidently evaporating them, and driving away a cloud of mist and volatile matter. This is when they can be seen. The comet is most gorgeous when it is near the sun, and as soon as it gets a reasonable distance away from him it is perfectly invisible.

The matter evaporated from the comet by the sun's heat does not return—it is lost to the comet; and hence, after a few such journeys, its volatile matter gets appreciably diminished, and so old-established periodic comets have no tails to speak of. But the new visitants, coming from the depths of space for the first time—these have great supplies of volatile matter, and these are they which show the most magnificent tails.

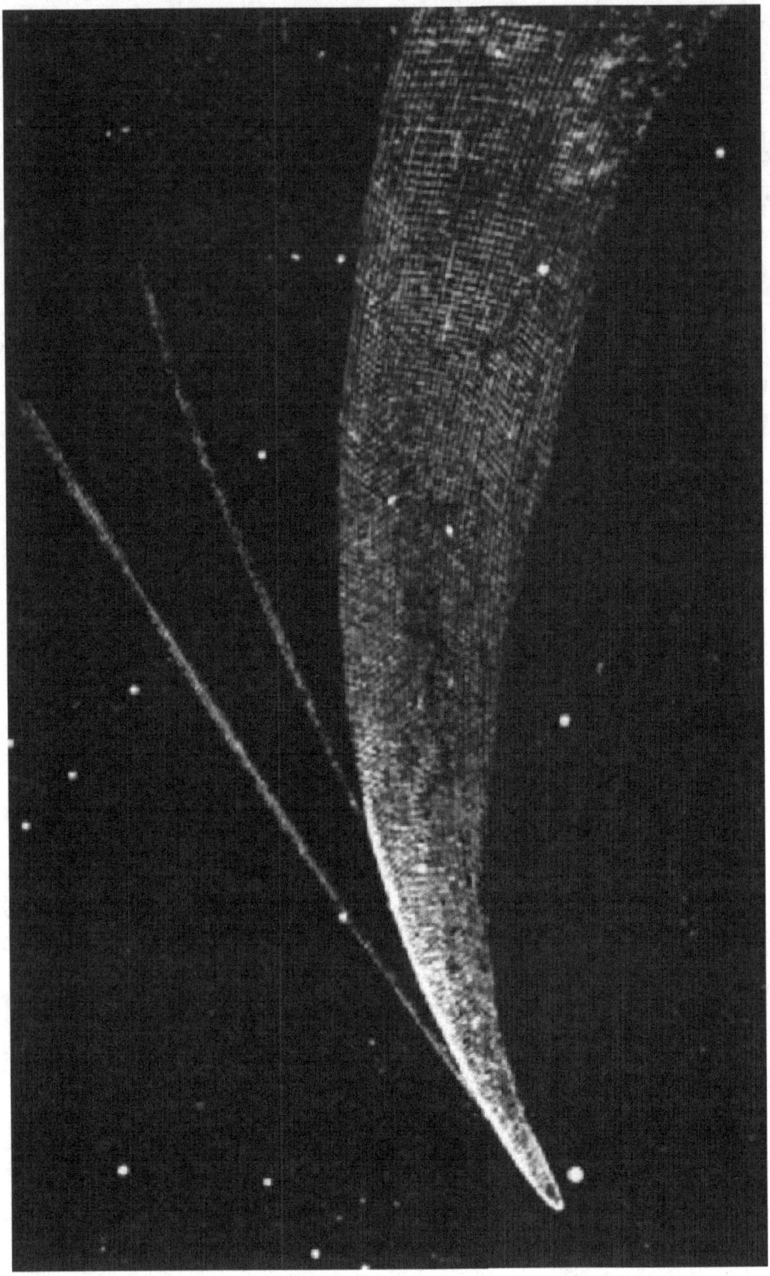

FIG. 101.—Head of Donati's comet of 1858.

The tail of a comet is always directed away from the sun as if it were repelled. To this rule there is no exception. It is suggested, and held as most probable, that the tail and sun are similarly electrified, and that the repulsion of the tail is electrical repulsion. Some great force is obviously at work to account for the enormous distance to which the tail is shot in a few hours. The pressure of the sun's light can do something, and is a force that must not be ignored when small particles are being dealt with. (Cf. *Modern Views of Electricity*, 2nd edition, p. 363.)

Now just think what analogies there are between comets and meteors. Both are bodies travelling in orbits round the sun, and both are mostly invisible, but both become visible to us under certain circumstances. Meteors become visible when they plunge into the extreme limits of our atmosphere. Comets become visible when they approach the sun. Is it possible that comets are large meteors which dip into the solar atmosphere, and are thus rendered conspicuously luminous? Certainly they do not dip into the actual main atmosphere of the sun, else they would be utterly destroyed; but it is possible that the sun has a faint trace of atmosphere extending far beyond this, and into this perhaps these meteors dip, and glow with the friction. The particles thrown off might be, also by friction, electrified; and the vaporous tail might be thus accounted for.

FIG. 102.—Halley's Comet.

Let us make this hypothesis provisionally—that comets are large meteors, or a compact swarm of meteors, which, coming near the sun, find a highly rarefied sort of atmosphere, in which they get heated and partly vaporized, just as ordinary meteorites do when they dip into the atmosphere of the earth. And let us see whether any facts bear out the analogy and justify the hypothesis.

I must tell you now the history of three bodies, and you will see that some intimate connection between comets and meteors is proved. The three bodies are known as, first, Encke's comet; second, Biela's comet; third, the November swarm of meteors.

Encke's comet (one of those discovered by Miss Herschel) is an insignificant-looking telescopic comet of small period, the orbit of which was well known, and which was carefully observed at each reappearance after Encke had calculated its orbit. It was the quickest of the comets, returning every 3½ years.

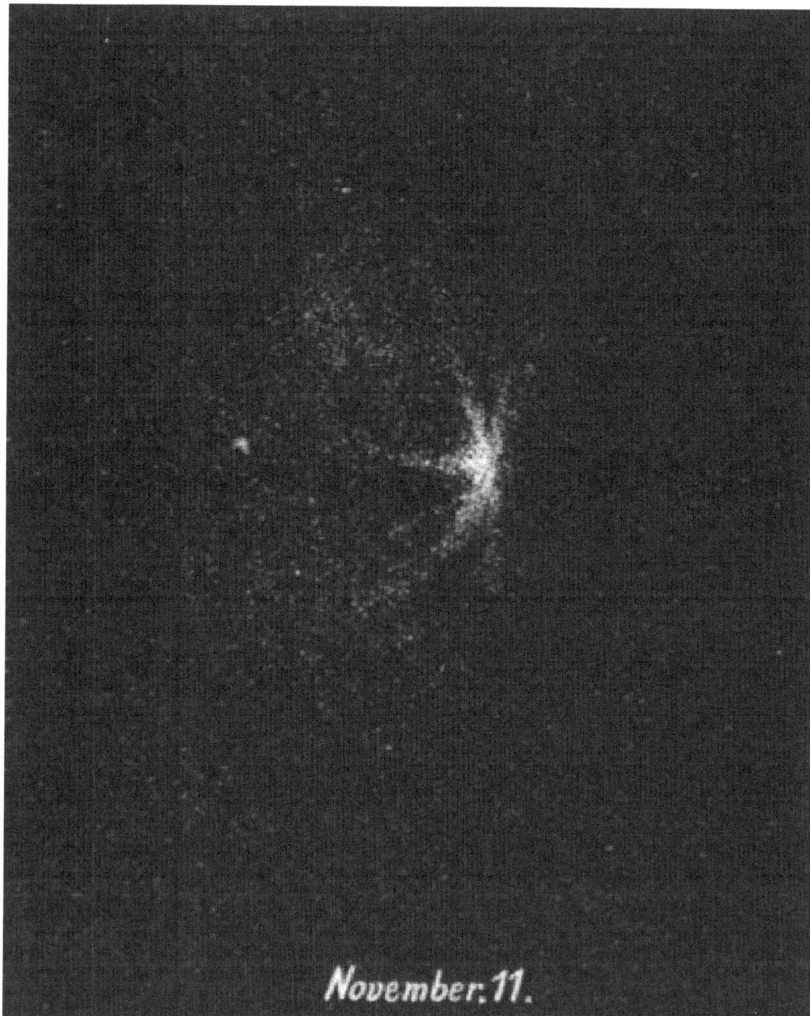

FIG. 103.—Encke's comet.

It was found, however, that its period was not quite constant; it kept on getting slightly shorter. The comet, in fact, returned to the sun slightly before its time. Now this effect is exactly what friction against a solar atmosphere would bring about. Every time it passed near the sun a little velocity would be rubbed out of it. But the velocity is that which carries it away, hence it would not go quite so far, and therefore would return a little sooner. Any revolving body subject to friction must revolve quicker and quicker, and get nearer and nearer its central body, until, if the process goes on long enough, it must drop upon its surface. This seems the kind of thing happening to Encke's comet. The effect is very small, and not thoroughly proved; but, so far as it goes, the evidence points to a greatly extended rare solar atmosphere, which rubs some energy out of it at every perihelion passage.

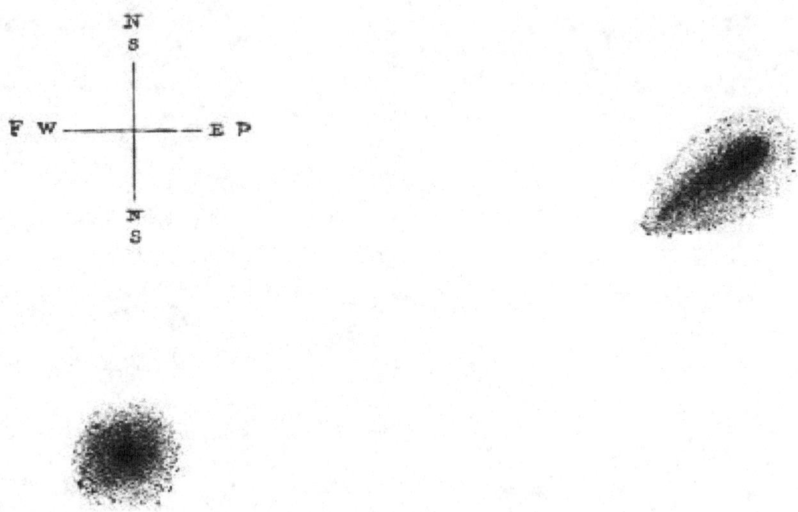

FIG. 104.—Biela's comet as last seen, in two portions.

Next, Biela's comet. This also was a well known and carefully observed telescopic comet, with a period of six years. In one of its distant excursions, it was calculated that it must pass very near Jupiter, and much curiosity was excited as to what would happen to it in consequence of the perturbation it must experience. As I have said, comets are only visible as they approach the sun, and a watch was kept for it about its appointed time. It was late, but it did ultimately arrive.

The singular thing about it, however, was that it was now double. It had apparently separated into two. This was in 1846. It was looked for again in 1852, and this time the components were further separated. Sometimes one was brighter, sometimes the other. Next time it ought to have come round no one could find either portion. The comet seemed to have wholly disappeared. It has never been seen since. It was then recorded and advertised as the missing comet.

But now comes the interesting part of the story. The orbit of this Biela comet was well known, and it was found that on a certain night in 1872 the earth would cross the orbit, and had some chance of encountering the comet. Not a very likely chance, because it need not be in that part of its orbit at the time; but it was suspected not to be far off—if still existent. Well, the night arrived, the earth did cross the orbit, and there was seen, not the comet, but a number of shooting-stars. Not one body, nor yet two, but a multitude of bodies—in fact, a swarm of meteors. Not a very great swarm, such as sometimes occurs, but still a quite noticeable one; and this shower of meteors is definitely recognized as flying along the track of Biela's comet. They are known as the Andromedes.

This observation has been generalized. Every cometary orbit is marked by a ring of meteoric stones travelling round it, and whenever a number of shooting-stars are seen quickly one after the other, it is an evidence that we are crossing the track of some comet. But suppose instead of only crossing the track of a comet we were to pass close to the comet itself, we should then expect to see an extraordinary swarm—a multitude of shooting-stars. Such phenomena have occurred. The most famous are those known as the November meteors, or Leonids.

This is the third of those bodies whose history I had to tell you. Professor H.A. Newton, of America, by examining ancient records arrived at the conclusion that the earth passed through a certain definite meteor shoal every thirty-three years. He found, in fact, that every thirty-three years an unusual flight of shooting-stars was witnessed in November, the earliest record being 599 A.D. Their last appearance had been in 1833, and he therefore predicted their return in 1866 or 1867. Sure enough, in November, 1866, they appeared; and many must remember seeing that glorious display. Although their hail was almost continuous, it is estimated that their average distance apart was thirty-five miles! Their radiant point was and always is in the constellation Leo, and hence their name Leonids.

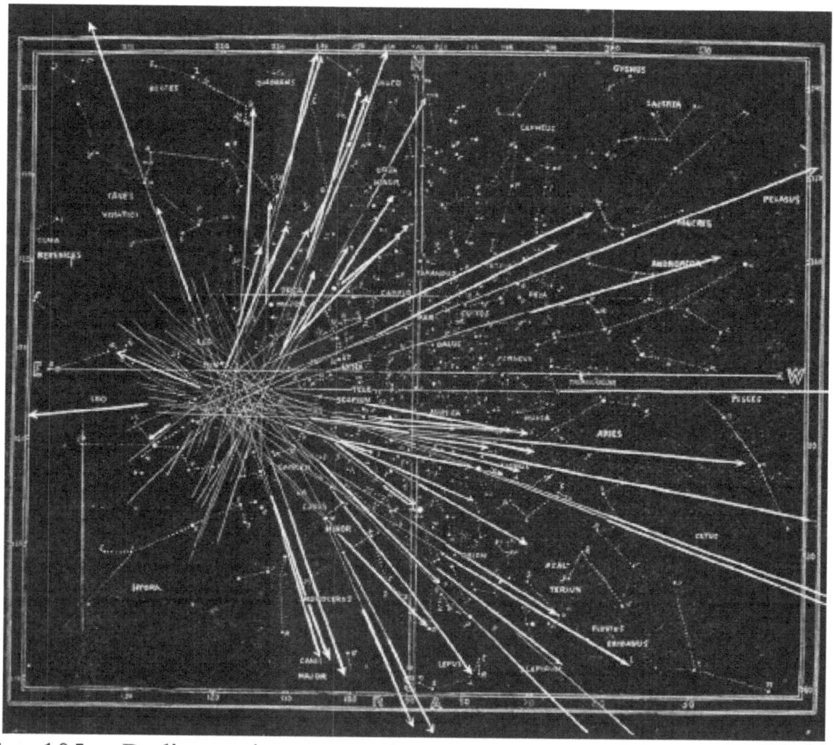

FIG. 105.—Radiant point perspective. The arrows represent a number of approximately parallel meteor-streaks foreshortened from a common vanishing-point.

A parallel stream fixed in space necessarily exhibits a definite aspect with reference to the fixed stars. Its aspect with respect to the earth will be very changeable, because of the rotation and revolution of that body, but its position with respect to constellations will be steady. Hence each meteor swarm, being a steady parallel stream of rushing masses, always strikes us from the same point in stellar space, and by this point (or radiant) it is identified and named.

The paths do not appear to us to be parallel, because of perspective: they seem to radiate and spread in all directions from a fixed centre like spokes, but all these diverging streaks are really parallel lines optically foreshortened by different amounts so as to produce the radiant impression.

The annexed diagram (Fig. 105) clearly illustrates the fact that the "radiant" is the vanishing point of a number of parallel lines.

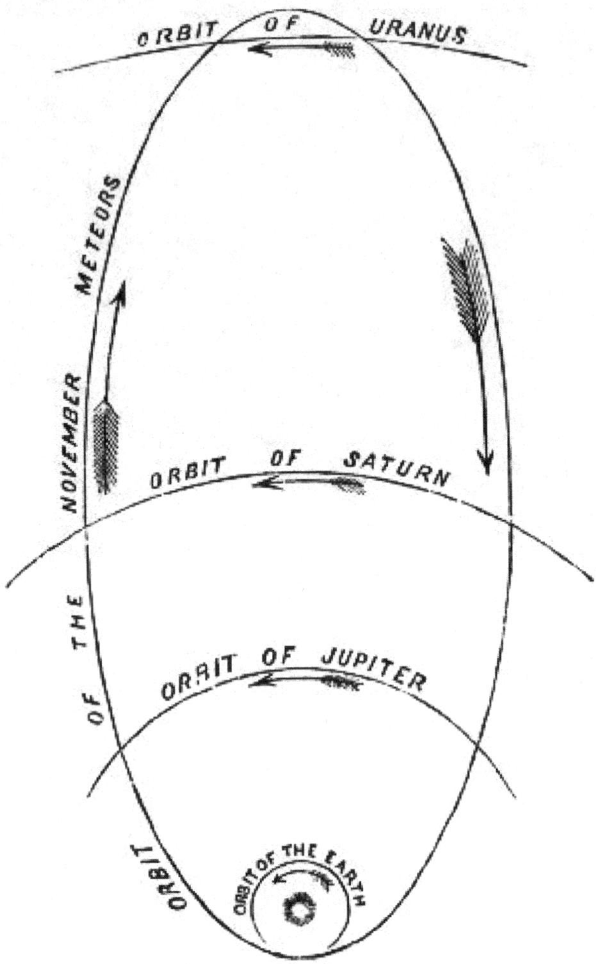

FIG. 106.—Orbit of November meteors.

This swarm is specially interesting to us from the fact that we cross its orbit every year. Its orbit and the earth's intersect. Every November we go through it, and hence every November we see a few stragglers of this immense swarm. The swarm itself takes thirty-three years on its revolution round the sun, and hence we only encounter it every thirty-three years.

The swarm is of immense size. In breadth it is such that the earth, flying nineteen miles a second, takes four or five hours to cross it, and this is therefore the time the display lasts. But in length it is far more enormous. The speed with which it travels is twenty-five miles a second, (for its orbit extends as far as Uranus, although by no means parabolic), and yet it takes more than a year to pass. Imagine a procession 200,000 miles broad, every individual rushing along at the rate of twenty-five miles every second, and the whole procession so long that it takes more than a year to pass. It is like a gigantic shoal of herrings swimming round and round the sun every thirty-three years, and travelling past the earth with that tremendous velocity of twenty-five miles a second. The earth dashes through the swarm and sweeps up myriads. Think of the countless numbers swept up by the whole earth in crossing such a shoal as that! But heaps more remain, and probably the millions which are destroyed every thirty-three years have not yet made any very important difference to the numbers still remaining.

The earth never misses this swarm. Every thirty-three years it is bound to pass through some part of them, for the shoal is so long that if the head is just missed one November the tail will be encountered next November. This is a plain and obvious result of its enormous length. It may be likened to a two-foot length of sewing silk swimming round and round an oval sixty feet in circumference. But, you will say, although the numbers are so great that destroying a few millions or so every thirty-three years makes but little difference to them, yet, if this process has been going on from all eternity, they ought to be all swept up. Granted; and no doubt the most ancient swarms have already all or nearly all been swept up.

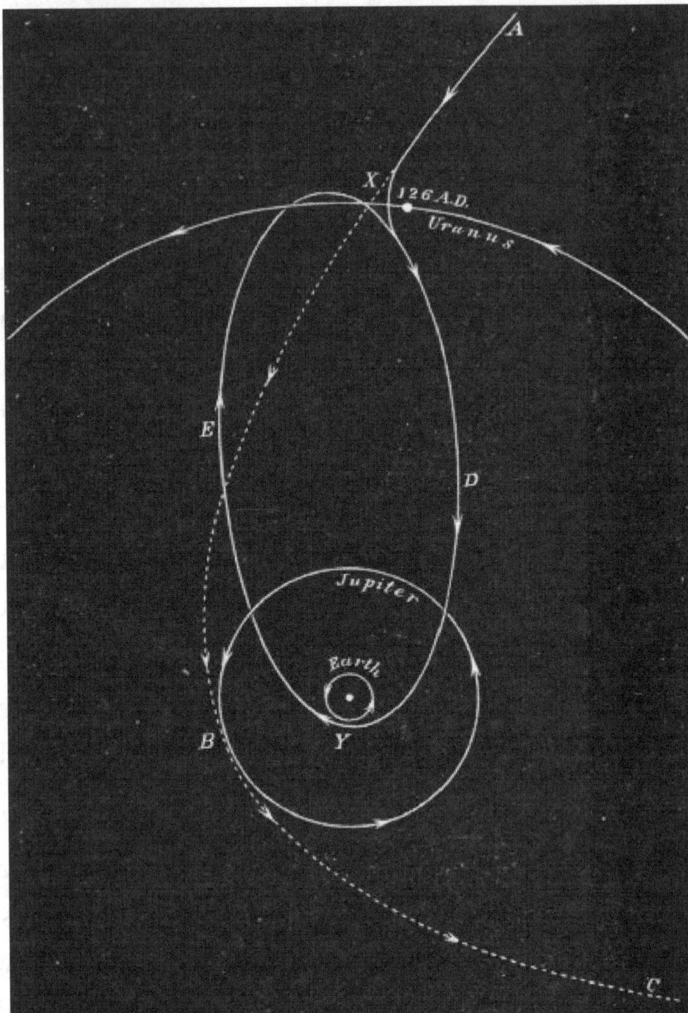

FIG. 107.—Orbit of November meteors; showing their probable parabolic orbit previous to 126 A.D., and its sudden conversion into an elliptic orbit by the violent perturbation caused by Uranus, which at that date occupied the position shown.

The August meteors, or Perseids, are an example. Every August we cross their path, and we have a small meteoric display radiating from the sword-hand of Perseus, but never specially more in one August than another. It would seem as if the main shoal has disappeared, and nothing is now left but the stragglers; or perhaps it is that the shoal has gradually become uniformly distributed all along the path. Anyhow, these August meteors are reckoned much more ancient members of the solar system than are the November meteors. The November meteors are believed to have entered the solar system in the year 126 A.D.

This may seem an extraordinary statement. It is not final, but it is based on the calculations of Leverrier—confirmed recently by Mr. Adams. A few moments will suffice to make the grounds of it clear. Leverrier calculated the orbit of the November meteors, and found them to be an oval extending beyond Uranus. It was perturbed by the outer planets near which it went, so that in past times it must have moved in a slightly different orbit. Calculating back to their past positions, it was found that in a certain year it must have gone very near to Uranus, and that by the perturbation of this planet its path had been completely changed. Originally it had in all probability been a comet, flying in a parabolic orbit towards the sun like many others. This one, encountering Uranus, was pulled to pieces as it were, and its orbit made elliptical as shown in Fig. 107. It was no longer free to escape and go away into the depths of space: it was enchained and made a member of the solar system. It also ceased to be a comet; it was degraded into a shoal of meteors.

This is believed to be the past history of this splendid swarm. Since its introduction to the solar system it has made 52 revolutions:

its next return is due in November, 1899, and I hope that it may occur in the English dusk, and (see Fig. 97) in a cloudless after-midnight sky, as it did in 1866.

NOTES FOR LECTURE XVII

The tide-generating force of one body on another is directly as the mass of the one body and inversely as the cube of the distance between them. Hence the moon is more effective in producing terrestrial tides than the sun.

The tidal wave directly produced by the moon in the open ocean is about 5 feet high, that produced by the sun is about 2 feet. Hence the average spring tide is to the average neap as about 7 to 3. The lunar tide varies between apogee and perigee from 4·3 to 5·9.

The solar tide varies between aphelion and perihelion from 1·9 to 2·1. Hence the highest spring tide is to the lowest neap as 5·9 + 2·1 is to 4·3 -2·1, or as 8 to 2·2.

The semi-synchronous oscillation of the Southern Ocean raises the magnitude of oceanic tides somewhat above these directly generated values.

Oceanic tides are true waves, not currents. Coast tides are currents. The momentum of the water, when the tidal wave breaks upon a continent and rushes up channels, raises coast tides to a much greater height—in some places up to 50 or 60 feet, or even more.

Early observed connections between moon and tides would be these:—

1st. Spring tides at new and full moon.

2nd. Average interval between tide and tide is half a lunar, not a solar, day—a lunar day being the interval between two successive returns of the moon to the meridian: 24 hours and 50 minutes.

3rd. The tides of a given place at new and full moon occur always at the same time of day whatever the season of the year.

LECTURE XVII

THE TIDES

PERSONS accustomed to make use of the Mersey landing-stages can hardly fail to have been struck with two obvious phenomena. One is that the gangways thereto are sometimes almost level, and at other times very steep; another is that the water often rushes past the stage rather violently, sometimes south towards Garston, sometimes north towards the sea. They observe, in fact, that the water has two periodic motions—one up and down, the other to and fro—a vertical and a horizontal motion. They may further observe, if they take the trouble, that a complete swing of the water, up and down, or to and fro, takes place about every twelve and a half hours; moreover, that soon after high and low water there is no current—the water is stationary, whereas about half-way between high and low it is rushing with maximum speed either up or down the river.

To both these motions of the water the name *tide* is given, and both are extremely important. Sailors usually pay most attention to the horizontal motion, and on charts you find the tide-races marked; and the places where there is but a small horizontal rush of the water are labelled "very little tide here." Landsmen, or, at any rate, such of the more philosophic sort as pay any attention to the matter at all, think most of the vertical motion of the water—its amount of rise and fall.

Dwellers in some low-lying districts in London are compelled to pay attention to the extra high tides of the Thames, because it is, or was, very liable to overflow its banks and inundate their basements.

Sailors, however, on nearing a port are also greatly affected by the time and amount of high water there, especially when they are in a big ship; and we know well enough how frequently Atlantic liners, after having accomplished their voyage with good speed, have to hang around for hours waiting till there is enough water to lift them over the Bar—that standing obstruction, one feels inclined to say disgrace, to the Liverpool harbour.

FIG. 108.—The Mersey

To us in Liverpool the tides are of supreme importance—upon them the very existence of the city depends—for without them Liverpool would not be a port. It may be familiar to many of you how this is, and yet it is a matter that cannot be passed over in silence. I will therefore call your attention to the Ordnance Survey of the estuaries of the Mersey and the Dee. You see first that there is a great tendency for sand-banks to accumulate all about this coast, from North Wales right away round to Southport. You see next that the port of Chester has been practically silted up by the deposits of sand in the wide-mouthed Dee, while the port of Liverpool remains open owing to the scouring action of the tide in its peculiarly shaped channel. Without the tides the Mersey would be a wretched dribble not much bigger than it is at Warrington. With them, this splendid basin is kept open, and a channel is cut of such depth that the *Great Eastern* easily rode in it in all states of the water.

The basin is filled with water every twelve hours through its narrow neck. The amount of water stored up in this basin at high tide I estimate as 600 million tons. All this quantity flows through the neck in six hours, and flows out again in the next six, scouring and cleansing and carrying mud and sand far out to sea. Just at present the currents set strongest on the Birkenhead side of the river, and accordingly a "Pluckington bank" unfortunately grows under the Liverpool stage. Should this tendency to silt up the gates of our docks increase, land can be reclaimed on the other side of the river between Tranmere and Rock Ferry, and an embankment made so as to deflect the water over Liverpool way, and give us a fairer proportion of the current. After passing New Brighton the water spreads out again to the left; its velocity forward diminishes; and after a few miles it has no power to cut away that sandbank known as the Bar. Should it be thought desirable to make it accomplish this, and sweep the Bar further out to sea into deeper water, it is probable that a rude training wall (say of old hulks, or other removable partial obstruction) on the west of Queen's Channel, arranged so as to check the spreading out over all this useless area, may be quite sufficient to retain the needed extra impetus in the water, perhaps even without choking up the useful old Rock Channel, through which smaller ships still find convenient exit.

Now, although the horizontal rush of the tide is necessary to our existence as a port, it does not follow that the accompanying rise and fall of the water is an unmixed blessing. To it is due the need for all the expensive arrangements of docks and gates wherewith to store up the high-level water. Quebec and New York are cities on such magnificent rivers that the current required to keep open channel is supplied without any tidal action, although Quebec is nearly 1,000 miles from the open ocean; and accordingly, Atlantic liners do not hover in mid-river and discharge passengers by tender, but they proceed straight to the side of the quays lining the river, or, as at New York, they dive into one of the pockets belonging to the company running the ship, and there discharge passengers and cargo without further trouble, and with no need for docks or gates. However, rivers like the St. Lawrence and the Hudson are the natural property of a gigantic continent; and we in England may be well contented with the possession of such tidal estuaries as the Mersey, the Thames, and the Humber. That by pertinacious dredging the citizens of Glasgow manage to get large ships right up their small river, the Clyde, to the quays of the town, is a remarkable fact, and redounds very highly to their credit.

We will now proceed to consider the connection existing between the horizontal rush of water and its vertical elevation, and ask, Which is cause and which is effect? Does the elevation of the ocean cause the tidal flow, or does the tidal flow cause the elevation? The answer is twofold: both statements are in some sense true. The prime cause of the tide is undoubtedly a vertical elevation of the ocean, a tidal wave or hump produced by the attraction of the moon. This hump as it passes the various channels opening into the ocean raises their level, and causes water to flow up them. But this simple oceanic tide, although the cause of all tide, is itself but a small affair. It seldom rises above six or seven feet, and tides on islands in mid-ocean have about this value or less. But the tides on our coasts are far greater than this—they rise twenty or thirty feet, or even fifty feet occasionally, at some places, as at Bristol. Why is this? The horizontal motion of the water gives it such an impetus or momentum that its motion far transcends that of the original impulse given to it, just as a push given to a pendulum may cause it to swing over a much greater arc than that through which the force acts. The inrushing water flowing up the English Channel or the Bristol Channel or St. George's Channel has such an impetus that it propels itself some twenty or thirty feet high before it has exhausted its momentum and begins to descend. In the Bristol Channel the gradual narrowing of the opening so much assists this action that the tides often rise forty feet, occasionally fifty feet, and rush still further up the Severn in a precipitous and extraordinary hill of water called "the bore."

Some places are subject to considerable rise and fall of water with very little horizontal flow; others possess strong tidal races, but very little elevation and depression. The effect observed at any given place entirely depends on whether the place has the general character of a terminus, or whether it lies *en route* to some great basin.

You must understand, then, that all tide takes its rise in the free and open ocean under the action of the moon. No ordinary-sized sea like the North Sea, or even the Mediterranean, is big enough for more than a just appreciable tide to be generated in it. The Pacific, the Atlantic, and the Southern Oceans are the great tidal reservoirs, and in them the tides of the earth are generated as low flat humps of gigantic area, though only a few feet high, oscillating up and down in the period of approximately twelve hours. The tides we, and other coast-possessing nations, experience are the overflow or back-wash of these oceanic humps, and I will now show you in what manner the great Atlantic tide-wave reaches the British Isles twice a day.

FIG. 109.—Co-tidal lines.

Fig. 109 shows the contour lines of the great wave as it rolls in east from the Atlantic, getting split by the Land's End and by Ireland into three portions; one of which rushes up the English Channel and through the Straits of Dover. Another rolls up the Irish Sea, with a minor offshoot up the Bristol Channel, and, curling round Anglesey, flows along the North Wales coast and fills Liverpool Bay and the Mersey. The third branch streams round the north coast of Ireland, past the Mull of Cantyre and Rathlin Island; part fills up the Firth of Clyde, while the rest flows south, and, swirling round the west side of the Isle of Man, helps the southern current to fill the Bay of Liverpool. The rest of the great wave impinges on the coast of Scotland, and, curling round it, fills up the North Sea right away to the Norway coast, and then flows down below Denmark, joining the southern and earlier arriving stream. The diagram I show you is a rough chart of cotidal lines, which I made out of the information contained in *Whitaker's Almanac*.

A place may thus be fed with tide by two distinct channels, and many curious phenomena occur in certain places from this cause. Thus it may happen that one channel is six hours longer than the other, in which case a flow will arrive by one at the same time as an ebb arrives by the other; and the result will be that the place will have hardly any tide at all, one tide interfering with and neutralizing the other. This is more markedly observed at other parts of the world than in the British Isles. Whenever a place is reached by two channels of different length, its tides are sure to be peculiar, and probably small.

Another cause of small tide is the way the wave surges to and fro in a channel. The tidal wave surging up the English Channel, for instance, gets largely reflected by the constriction at Dover, and so a crest surges back again, as we may see waves reflected in a long trough or tilted bath. The result is that Southampton has two high tides rapidly succeeding one another, and for three hours the high-water level varies but slightly—a fact of evident convenience to the port.

Places on a nodal line, so to speak, about the middle of the length of the channel, have a minimum of rise and fall, though the water rushes past them first violently up towards Dover, where the rise is considerable, and then back again towards the ocean. At Portland, for instance, the total rise and fall is very small: it is practically on a node. Yarmouth, again, is near a less marked node in the North Sea, where stationary waves likewise surge to and fro, and accordingly the tidal rise and fall at Yarmouth is only about five feet (varying from four and a half to six), whereas at London it is twenty or thirty feet, and at Flamborough Head or Leith it is from twelve to sixteen feet.

It is generally supposed that water never flows up-hill, but in these cases of oscillation it flows up-hill for three hours together. The water is rushing up the English Channel towards Dover long after it is highest at the Dover end; it goes on piling itself up, until its momentum is checked by the pressure, and then it surges back. It behaves, in fact, very like the bob of a pendulum, which rises against gravity at every quarter swing.

To get a very large tide, the place ought to be directly accessible by a long sweep of a channel to the open ocean, and if it is situate on a gradually converging opening, the ebb and flow may be enormous. The Severn is the best example of this on the British Isles; but the largest tides in the world are found, I believe, in the Bay of Fundy, on the coast of North America, where they sometimes rise one hundred and twenty feet. Excessive or extra tides may be produced occasionally in any place by the propelling force of a high wind driving the water towards the shore; also by a low barometer, *i.e.* by a local decrease in the pressure of the air.

Well, now, leaving these topographical details concerning tides, which we see to be due to great oceanic humps (great in area that is, though small in height), let us proceed to ask what causes these humps; and if it be the moon that does it, how does it do it?

The statement that the moon causes the tides sounds at first rather an absurdity, and a mere popular superstition. Galileo chaffed Kepler for believing it. Who it was that discovered the connection between moon and tides we know not—probably it is a thing which has been several times rediscovered by observant sailors or coast-dwellers—and it is certainly a very ancient piece of information.

Probably the first connection observed was that about full moon and about new moon the tides are extra high, being called spring tides, whereas about half-moon the tides are much less, and are called neap tides. The word spring in this connection has no reference to the season of the year; except that both words probably represent the same idea of energetic uprising or upspringing, while the word neap comes from nip, and means pinched, scanty, nipped tide.

The next connection likely to be observed would be that the interval between two day tides was not exactly a solar day of twenty-four hours, but a lunar day of fifty minutes longer. For by reason of the moon's monthly motion it lags behind the sun about fifty minutes a day, and the tides do the same, and so perpetually occur later and later, about fifty minutes a day later, or 12 hours and 25 minutes on the average between tide and tide.

A third and still more striking connection was also discovered by some of the ancient great navigators and philosophers—viz. that the time of high water at a given place at full moon is always the same, or very nearly so. In other words, the highest or spring tides always occur nearly at the same time of day at a given place. For instance, at Liverpool this time is noon and midnight. London is about two hours and a half later. Each port has its own time for receiving a given tide, and the time is called the "establishment" of the port. Look out a day when the moon is full, and you will find the Liverpool high tide occurs at half-past eleven, or close upon it. The same happens when the moon is new. A day after full or new moon the spring tides rise to their highest, and these extra high tides always occur in Liverpool at noon and at midnight, whatever the season of the year. About the equinoxes they are liable to be extraordinarily high. The extra low tides here are therefore at 6 a.m. and 6 p.m., and the 6 p.m. low tide is a nuisance to the river steamers. The spring tides at London are highest about half-past two.

It is, therefore, quite clear that the moon has to do with the tides. It and the sun together are, in fact, the whole cause of them; and the mode in which these bodies act by gravitative attraction was first made out and explained in remarkably full detail by Sir Isaac Newton. You will find his account of the tides in the second and third books of the *Principia*; and though the theory does not occupy more than a few pages of that immortal work, he succeeds not only in explaining the local tidal peculiarities, much as I have done to-night, but also in calculating the approximate height of mid-ocean solar tide; and from the observed lunar tide he shows how to determine the then quite unknown mass of the moon. This was a quite extraordinary achievement, the difficulty of which it is not easy for a person unused to similar discussions fully to appreciate.

It is, indeed, but a small part of what Newton accomplished, but by itself it is sufficient to confer immortality upon any ordinary philosopher, and to place him in a front rank.

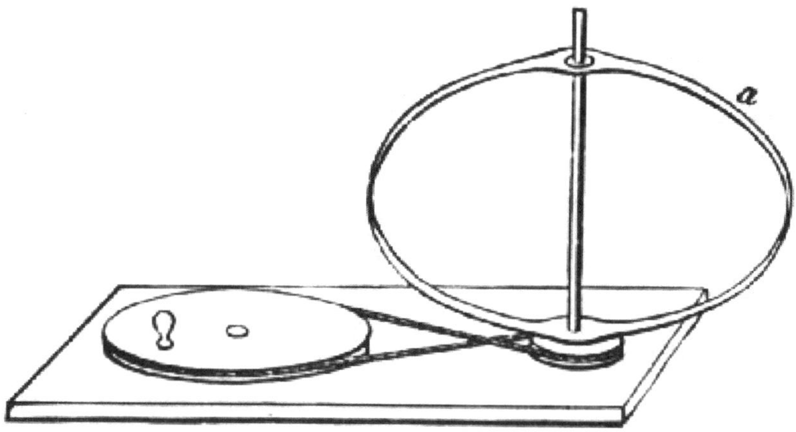

FIG. 110.—Whirling earth model.

To make intelligible Newton's theory of the tides, I must not attempt to go into too great detail. I will consider only the salient points. First, you know that every mass of matter attracts every other piece of matter; second, that the moon revolves round the earth, or rather that the earth and moon revolve round their common centre of gravity once a month; third, that the earth spins on its own axis once a day; fourth, that when a thing is whirled round, it tends to fly out from the centre and requires a force to hold it in. These are the principles involved. You can whirl a bucket full of water vertically round without spilling it. Make an elastic globe rotate, and it bulges out into an oblate or orange shape; as illustrated by the model shown in Fig. 110. This is exactly what the earth does, and Newton calculated the bulging of it as fourteen miles all round the equator. Make an elastic globe revolve round a fixed centre outside itself, and it gets pulled into a prolate or lemon shape; the simplest illustrative experiment is to attach a string to an elastic bag or football full of water, and whirl it round and round. Its prolateness is readily visible.

Now consider the earth and moon revolving round each other like a man whirling a child round. The child travels furthest, but the man cannot merely rotate, he leans back and thus also describes a small circle: so does the earth; it revolves round the common centre of gravity of earth and moon (*cf.* p. 212). This is a vital point in the comprehension of the tides: the earth's centre is not at rest, but is being whirled round by the moon, in a circle about 1⁄80 as big as the circle which the moon describes, because the earth weighs eighty times as much as the moon. The effect of the revolution is to make both bodies slightly protrude in the direction of the line joining them; they become slightly "prolate" as it is called—that is, lemon-shaped. Illustrating still by the man and child, the child's legs fly outwards so that he is elongated in the direction of a radius; the man's coat-tails fly out too, so that he too is similarly though less elongated. These elongations or protuberances constitute the tides.

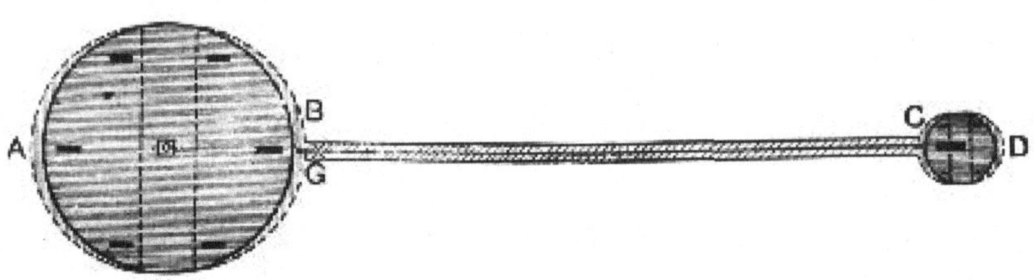

FIG. 111.—Earth and moon model, illustrating the production of statical or "equilibrium" tides when the whole is whirled about the point G.

Fig. 111 shows a model to illustrate the mechanism. A couple of cardboard disks (to represent globes of course), one four times the diameter of the other, and each loaded so as to have about the correct earth-moon ratio of weights, are fixed at either end of a long stick, and they balance about a certain point, which is their common centre of gravity. For convenience this point is taken a trifle too far out from the centre of the earth—that is, just beyond its surface. Through the balancing point G a bradawl is stuck, and on that as pivot the whole readily revolves. Now, behind the circular disks, you see, are four pieces of card of appropriate shape, which are able to slide out under proper forces. They are shown dotted in the figure, and are lettered A, B, C, D. The inner pair, B and C, are attached to each other by a bit of string, which has to typify the attraction of gravitation; the outer pair, A and D, are not attached to anything, but have a certain amount of play against friction in slots parallel to the length of the stick. The moon-disk is also slotted, so a small amount of motion is possible to it along the stick or bar. These things being so arranged, and the protuberant pieces of card being all pushed home, so that they are hidden behind their respective disks, the whole is spun rapidly round the centre of gravity, G. The result of a brief spin is to make A and D fly out by centrifugal force and show, as in the figure; while the moon, flying out too in its slot, tightens up the string, which causes B and C to be pulled out too. Thus all four high tides are produced, two on the earth and two on the moon, A and D being caused by centrifugal force, B and C by the attraction of gravitation. Each disk has become prolate in the same sort of fashion as yielding globes do. Of course the fluid ocean takes this shape more easily and more completely than the solid earth can, and so here are the very oceanic humps we have been talking about, and about three feet high (Fig. 112). If there were a sea on the *moon*, its humps would be a good deal bigger; but there probably is no sea there, and if there were, the earth's tides are more interesting to us, at any rate to begin with.

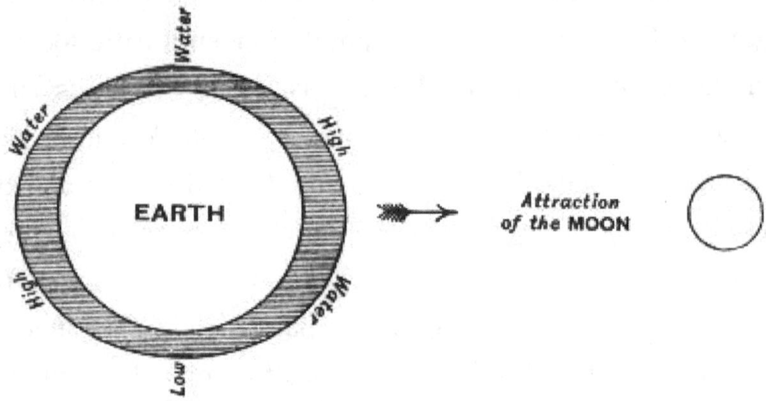

FIG. 112.—Earth and moon (earth's rotation neglected).

The humps as so far treated are always protruding in the earth-moon line, and are stationary. But now we have to remember that the earth is spinning inside them. It is not easy to see what precise effect this spin will have upon the humps, even if the world were covered with a uniform ocean; but we can see at any rate that however much they may get displaced, and they do get displaced a good deal, they cannot possibly be carried round and round. The whole explanation we have given of their causes shows that they must maintain some steady aspect with respect to the moon—in other words, they must remain stationary as the earth spins round. Not that the same identical water remains stationary, for in that case it would have to be dragged over the earth's equator at the rate of 1,000 miles an hour, but the hump or wave-crest remains stationary. It is a true wave, or form only, and consists of continuously changing individual particles. The same is true of all waves, except breaking ones.

Given, then, these stationary humps and the earth spinning on its axis, we see that a given place on the earth will be carried round and round, now past a hump, and six hours later past a depression: another six hours and it will be at the antipodal hump, and so on. Thus every six hours we shall travel from the region in space where the water is high to the region where it is low; and ignoring our own motion we shall say that the sea first rises and then falls; and so, with respect to the place, it does. Thus the succession of high and low water, and the two high tides every twenty-four hours, are easily understood in their easiest and most elementary aspect. A more complete account of the matter it will be wisest not to attempt: suffice it to say that the difficulties soon become formidable when the inertia of the water, its natural time of oscillation, the varying obliquity of the moon to the ecliptic, its varying distance, and the disturbing action of the sun are taken into consideration. When all these things are included, the problem becomes to ordinary minds overwhelming. A great many of these difficulties were successfully attacked by Laplace. Others remained for modern philosophers, among whom are Sir George Airy, Sir William Thomson, and Professor George Darwin.

I may just mention that the main and simplest effect of including the inertia or momentum of the water is to dislocate the obvious and simple connexion between high water and high moon; inertia always tends to make an effect differ in phase by a quarter period from the cause producing it, as may be illustrated by a swinging pendulum. Hence high water is not to be expected when the tide-raising force is a maximum, but six hours later; so that, considering inertia and neglecting friction, there would be low water under the moon. Including friction, something nearer the equilibrium state of things occurs. With *sufficient* friction the motion becomes dead-beat again, *i.e.* follows closely the force that causes it.

Returning to the elementary discussion, we see that the rotation of the earth with respect to the humps will not be performed in exactly twenty-four hours, because the humps are travelling slowly after the moon, and will complete a revolution in a month in the same direction as the earth is rotating. Hence a place on the earth has to catch them up, and so each high tide arrives later and later each day—roughly speaking, an hour later for each day tide; not by any means a constant interval, because of superposed disturbances not here mentioned, but on the average about fifty minutes.

We see, then, that as a result of all this we get a pair of humps travelling all over the surface of the earth, about once a day. If the earth were all ocean (and in the southern hemisphere it is nearly all ocean), then they would go travelling across the earth, tidal waves three feet high, and constituting the mid-ocean tides. But in the northern hemisphere they can only thus journey a little way without striking land. As the moon rises at a place on the east shores of the Atlantic, for instance, the waters begin to flow in towards this place, or the tide begins to rise. This goes on till the moon is overhead and for some time afterwards, when the tide is at its highest. The hump then follows the moon in its apparent journey across to America, and there precipitates itself upon the coast, rushing up all the channels, and constituting the land tide. At the same time, the water is dragged away from the east shores, and so *our* tide is at its lowest. The same thing repeats itself in a little more than twelve hours again, when the other hump passes over the Atlantic, as the moon journeys beneath the earth, and so on every day.

In the free Southern Ocean, where land obstruction is comparatively absent, the water gets up a considerable swing by reason of its accumulated momentum, and this modifies and increases the open ocean tides there. Also for some reason, I suppose because of the natural time of swing of the water, one of the humps is there usually much larger than the other; and so places in the Indian and other offshoots of the Southern Ocean get their really high tide only once every twenty-four hours. These southern tides are in fact much more complicated than those the British Isles receive. Ours are singularly simple. No doubt some trace of the influence of the Southern Ocean is felt in the North Atlantic, but any ocean extending over 90° of longitude is big enough to have its own tides generated; and I imagine that the main tides we feel are thus produced on the spot, and that they are simple because the damping-out being vigorous, and accumulated effects small, we feel the tide-producing forces more directly. But for authoritative statements on tides, other books must be read. I have thought, and still think, it best in an elementary exposition to begin by a consideration of the tide-generating forces as if they acted on a non-rotating earth. It is the tide generating forces, and not the tides themselves, that are really represented in Figs. 112 and 114. The rotation of the earth then comes in as a disturbing cause. A more complete exposition would begin with the rotating earth, and would superpose the attraction of the moon as a disturbing cause, treating it as a problem in planetary perturbation, the ocean being a sort of satellite of the earth. This treatment, introducing inertia but ignoring friction and land obstruction, gives low water in the line of pull, and high water at right angles, or where the pull is zero; in the same sort of way as a pendulum bob is highest where most force is pulling it down, and lowest where no force is acting on it. For a clear treatment of the tides as due to the perturbing forces of sun and moon, see a little book by Mr. T.K. Abbott of Trinity College, Dublin. (Longman.)

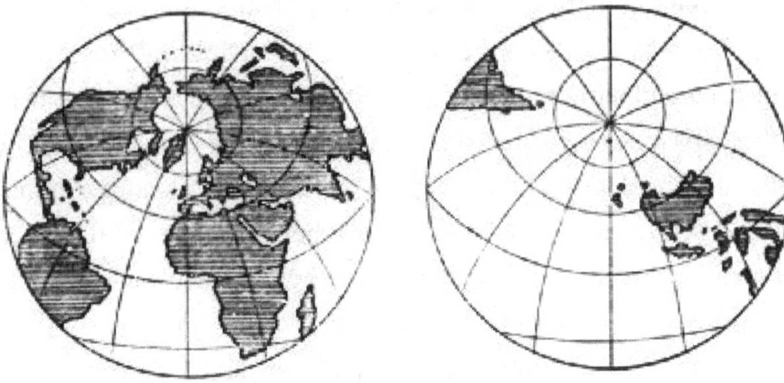

FIG. 113.—Maps showing how comparatively free from land obstruction the ocean in the Southern Hemisphere is.

If the moon were the only body that swung the earth round, this is all that need be said in an elementary treatment; but it is not the only one. The moon swings the earth round once a month, the sun swings it round once a year. The circle of swing is bigger, but the speed is so much slower that the protuberance produced is only one-third of that caused by the monthly whirl; *i.e.* the simple solar tide in the open sea, without taking momentum into account, is but a little more than a foot high, while the simple lunar tide is about three feet. When the two agree, we get a spring tide of four feet; when they oppose each other, we get a neap tide of only two feet. They assist each other at full moon and at new moon. At half-moon they oppose each other. So we have spring tides regularly once a fortnight, with neap tides in between.

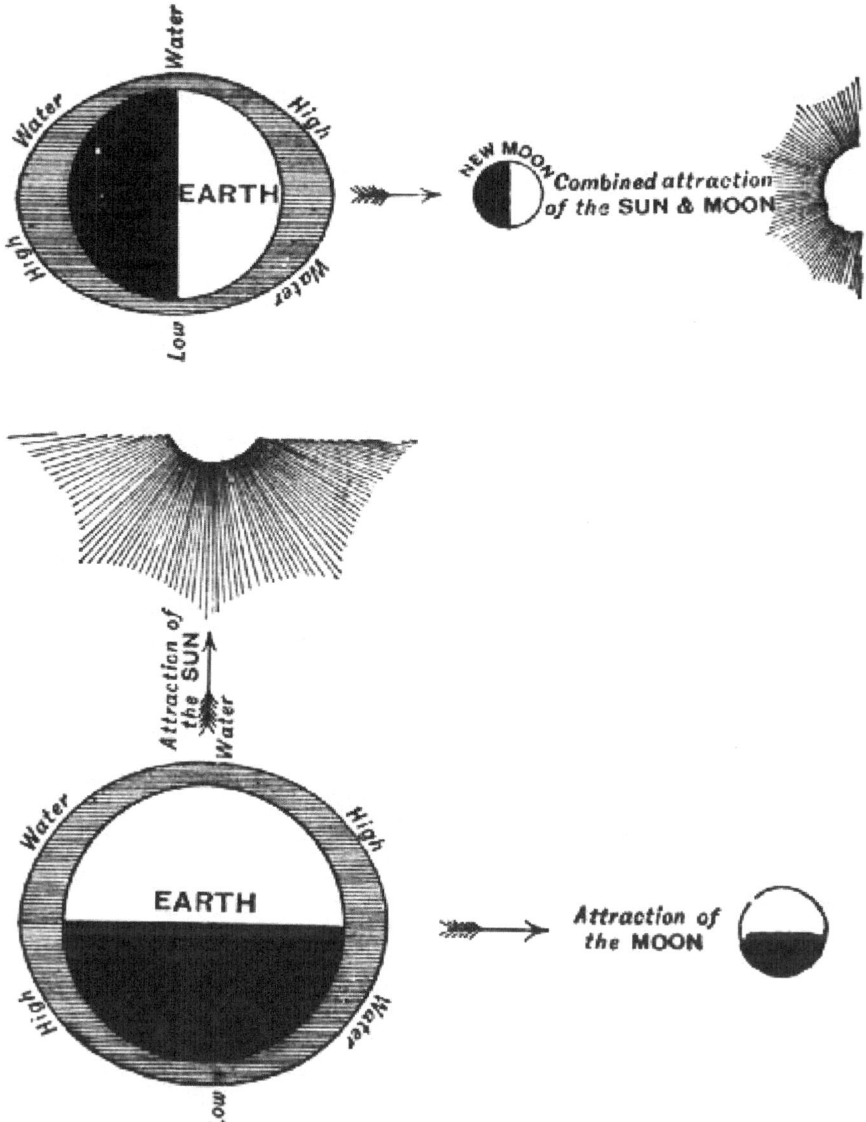

FIG. 114.—Spring and neap tides.

Fig. 114 gives the customary diagrams to illustrate these simple things. You see that when the moon and sun act at right angles (*i.e.* at every half-moon), the high tides of one coincide with the low tides of the other; and so, as a place is carried round by the earth's rotation, it always finds either solar or else lunar high water, and only experiences the difference of their two effects. Whereas, when the sun and moon act in the same line (as they do at new and full moon), their high and low tides coincide, and a place feels their effects added together. The tide then rises extra high and falls extra low.

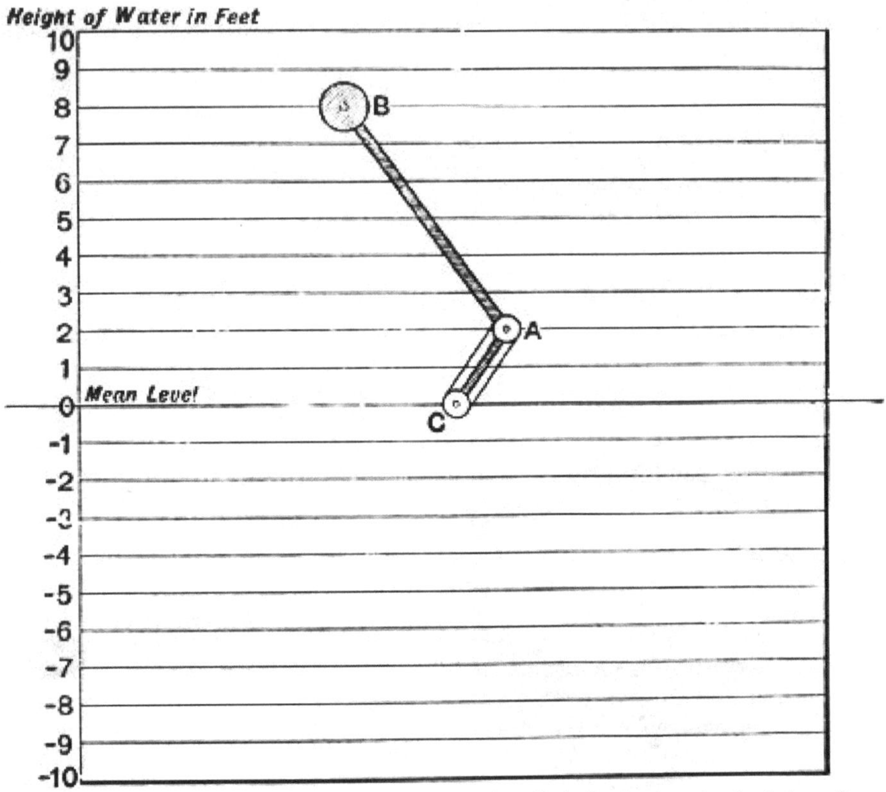

FIG. 115.—Tidal clock. The position of the disk B shows the height of the tide. The tide represented is a nearly high tide eight feet above mean level.

Utilizing these principles, a very elementary form of tidal-clock, or tide-predicter, can be made, and for an open coast station it really would not give the tides so very badly. It consists of a sort of clock face with two hands, one nearly three times as long as the other. The short hand, CA, should revolve round C once in twelve hours, and the vertical height of its end A represents the height of the solar tide on the scale of horizontal lines ruled across the face of the clock. The long hand, AB, should revolve round A once in twelve hours and twenty-five minutes, and the height of its end B (if A were fixed on the zero line) would represent the lunar tide. The two revolutions are made to occur together, either by means of a link-work parallelogram, or, what is better in practice, by a string and pulleys, as shown; and the height of the end point, B, of the third side or resultant, CB, read off on a scale of horizontal parallel lines behind, represents the combination or actual tide at the place. Every fortnight the two will agree, and you will get spring tides of maximum height CA + AB; every other fortnight the two will oppose, and you will get neap tides of maximum height CA-AB.

Such a clock, if set properly and driven in the ordinary way, would then roughly indicate the state of the tide whenever you chose to look at it and read the height of its indicating point. It would not indeed be very accurate, especially for such an inclosed station as Liverpool is, and that is probably why they are not made. A great number of disturbances, some astronomical, some terrestrial, have to be taken into account in the complete theory. It is not an easy matter to do this, but it can be, and has been, done; and a tide-predicter has not only been constructed, but two of them are in regular work, predicting the tides for years hence—one, the property of the Indian Government, for coast stations of India; the other for various British and foreign stations, wherever the necessary preliminary observations have been made. These machines are the invention of Sir William Thomson. The tide-tables for Indian ports are now always made by means of them.

FIG. 116.—Sir William Thomson (Lord Kelvin).

FIG. 117.—Tide-gauge for recording local tides, a pencil moved up and down by a float writes on a drum driven by clockwork.

The first thing to be done by any port which wishes its tides to be predicted is to set up a tide-gauge, or automatic recorder, and keep it working for a year or two. The tide-gauge is easy enough to understand: it marks the height of the tide at every instant by an irregular curved line like a barometer chart (Fig. 117). These observational curves so obtained have next to be fed into a fearfully complex machine, which it would take a whole lecture to make even partially intelligible, but Fig. 118 shows its aspect. It consists of ten integrating machines in a row, coupled up and working together. This is the "harmonic analyzer," and the result of passing the curve through this machine is to give you all the constituents of which it is built up, viz. the lunar tide, the solar tide, and eight of the sub-tides or disturbances. These ten values are then set off into a third machine, the tide-predicter proper. The general mode of action of this machine is not difficult to understand. It consists of a string wound over and under a set of pulleys, which are each set on an excentric, so as to have an up-and-down motion. These up-and-down motions are all different, and there are ten of these movable pulleys, which by their respective excursions represent the lunar tide, the solar tide, and the eight disturbances already analyzed out of the tide-gauge curve by the harmonic analyzer. One end of the string is fixed, the other carries a pencil which writes a trace on a revolving drum of paper—a trace which represents the combined motion of all the pulleys, and so predicts the exact height of the tide at the place, at any future time you like. The machine can be turned quite quickly, so that a year's tides can be run off with every detail in about half-an-hour. This is the easiest part of the operation. Nothing has to be done but to keep it supplied with paper and pencil, and turn a handle as if it were a coffee-mill instead of a tide-mill. (Figs. 119 and 120.)

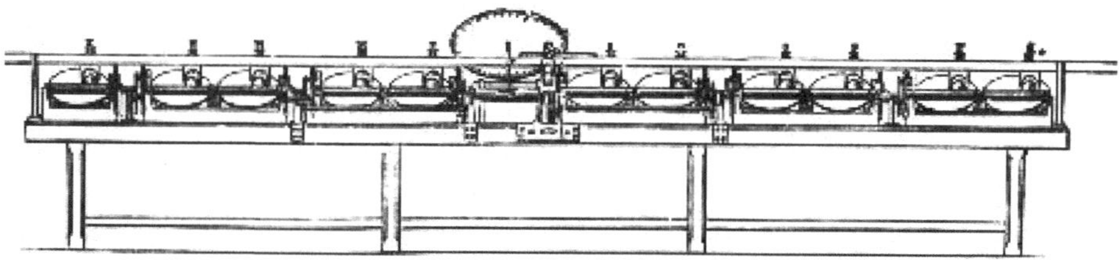

FIG. 118.—Harmonic analyzer; for analyzing out the constituents from a set of observational curves.

My subject is not half exhausted. I might go on to discuss the question of tidal energy—whether it can be ever utilized for industrial purposes; and also the very interesting question whence it comes. Tidal energy is almost the only terrestrial form of energy that does not directly or indirectly come from the sun. The energy of tides is now known to be obtained at the expense of the earth's rotation; and accordingly our day must be slowly, very slowly, lengthening. The tides of past ages have destroyed the moon's rotation, and so it always turns the same face to us. There is every reason to believe that in geologic ages the moon was nearer to us than it is now, and that accordingly our tides were then far more violent, rising some hundreds of feet instead of twenty or thirty, and sweeping every six hours right over the face of a country, ploughing down hills, denuding rocks, and producing a copious sedimentary deposit.

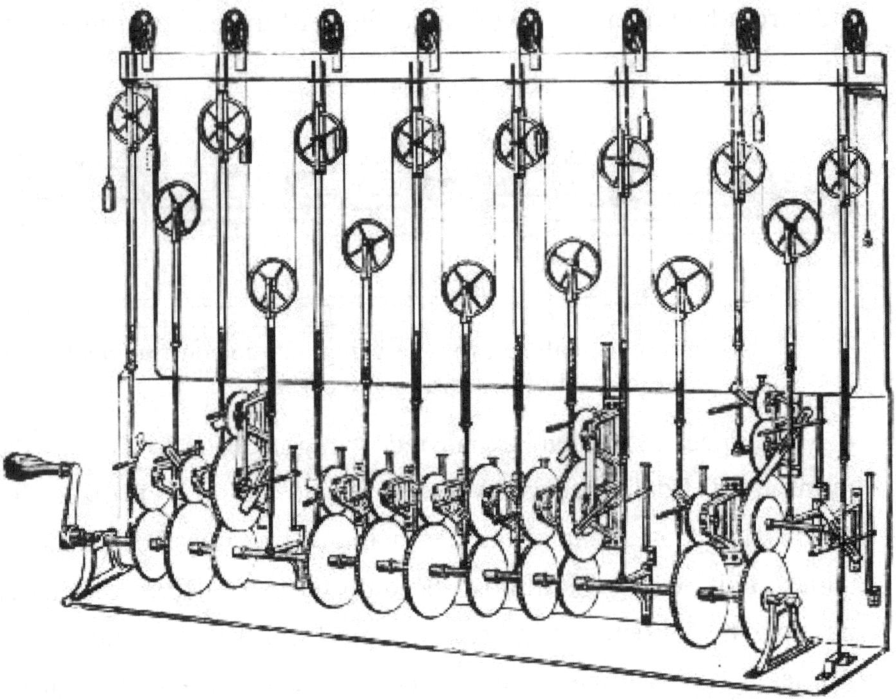

FIG. 119.—Tide-predicter, for combining the ascertained constituents into a tidal curve for the future.

In thus discovering the probable violent tides of past ages, astronomy has, within the last few years, presented geology with the most powerful denuding agent known; and the study of the earth's past history cannot fail to be greatly affected by the modern study of the intricate and refined conditions attending prolonged tidal action on incompletely rigid bodies. [Read on this point the last chapter of Sir R. Ball's *Story of the Heavens*.]

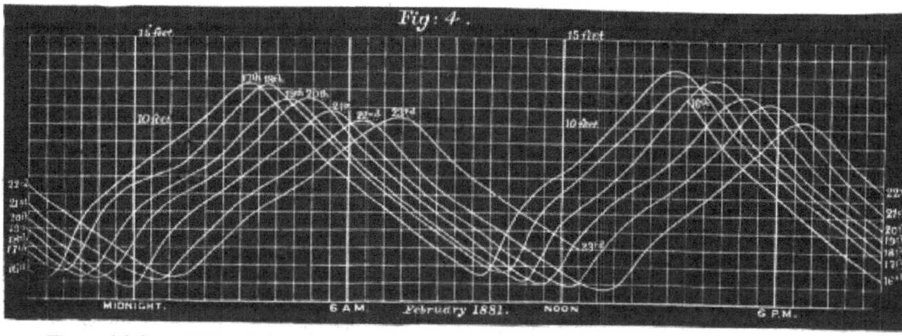

FIG. 120.—Weekly sheet of curves. Tides for successive days are predicted on the same sheet of paper, to economise space.

I might also point out that the magnitude of our terrestrial tides enables us to answer the question as to the internal fluidity of the earth. It used to be thought that the earth's crust was comparatively thin, and that it contained a molten interior. We now know that this is not the case. The interior of the earth is hot indeed, but it is not fluid. Or at least, if it be fluid, the amount of fluid is but very small compared with the thickness of the unyielding crust. All these, and a number of other most interesting questions, fringe the subject of the tides; the theoretical study of which, started by Newton, has developed, and is destined in the future to further develop, into one of the most gigantic and absorbing investigations—having to do with the stability or instability of solar systems, and with the construction and decay of universes.

These theories are the work of pioneers now living, whose biographies it is therefore unsuitable for us to discuss, nor shall I constantly mention their names. But Helmholtz, and Thomson, are household words, and you well know that in them and their disciples the race of Pioneers maintains its ancient glory.

NOTES FOR LECTURE XVIII

Tides are due to incomplete rigidity of bodies revolving round each other under the action of gravitation, and at the same time spinning on their axes.

Two spheres revolving round each other can only remain spherical if rigid; if at all plastic they become prolate. If either rotate on its axis, in the same or nearly the same plane as it revolves, that one is necessarily subject to tides.

The axial rotation tends to carry the humps with it, but the pull of the other body keeps them from moving much. Hence the rotation takes place against a pull, and is therefore more or less checked and retarded. This is the theory of Von Helmholtz.

The attracting force between two such bodies is no longer *exactly* towards the centre of revolution, and therefore Kepler's second law is no longer precisely obeyed: the rate of description of areas is subject to slight acceleration. The effect of this tangential force acting on the tide-compelling body is gradually to increase its distance from the other body.

Applying these statements to the earth and moon, we see that tidal energy is produced at the expense of the earth's rotation, and that the length of the day is thereby slowly increasing. Also that the moon's rotation relative to the earth has been destroyed by past tidal action in it (the only residue of ancient lunar rotation now being a scarcely perceptible libration), so that it turns always the same face towards us. Moreover, that its distance from the earth is steadily increasing. This last is the theory of Professor G.H. Darwin.

Long ago the moon must therefore have been much nearer the earth, and the day was much shorter. The tides were then far more violent.

Halving the distance would make them eight times as high; quartering it would increase them sixty-four-fold. A most powerful geological denuding agent. Trade winds and storms were also more violent.

If ever the moon were close to the earth, it would have to revolve round it in about three hours. If the earth rotated on its axis in three hours, when fluid or pasty, it would be unstable, and begin to separate a portion of itself as a kind of bud, which might then get detached and gradually pushed away by the violent tidal action. Hence it is possible that this is the history of the moon. If so, it is probably an exceptional history. The planets were not formed from the sun in this way.

Mars' moons revolve round him more quickly than the planet rotates: hence with them the process is inverted, and they must be approaching him and may some day crash along his surface. The inner moon is now about 4,000 miles away, and revolves in 7½. It appears to be about 20 miles in diameter, and weighs therefore, if composed of rock, 40 billion tons. Mars rotates in 24½ hours.

A similar fate may *possibly* await our moon ages hence—by reason of the action of terrestrial tides produced by the sun.

LECTURE XVIII

THE TIDES, AND PLANETARY EVOLUTION

In the last lecture we considered the local peculiarities of the tides, the way in which they were formed in open ocean under the action of the moon and the sun, and also the means by which their heights and times could be calculated and predicted years beforehand. Towards the end I stated that the subject was very far from being exhausted, and enumerated some of the large and interesting questions which had been left untouched. It is with some of these questions that I propose now to deal.

I must begin by reminding you of certain well-known facts, a knowledge of which I may safely assume.

And first we must remind ourselves of the fact that almost all the rocks which form the accessible crust of the earth were deposited by the agency of water. Nearly all are arranged in regular strata, and are composed of pulverized materials—materials ground down from pre-existing rocks by some denuding and grinding action. They nearly all contain vestiges of ancient life embedded in them, and these vestiges are mainly of marine origin. The strata which were once horizontal are now so no longer—they have been tilted and upheaved, bent and distorted, in many places. Some of them again have been metamorphosed by fire, so that their organic remains have been destroyed, and the traces of their aqueous origin almost obliterated. But still, to the eye of the geologist, all are of aqueous or sedimentary origin: roughly speaking, one may say they were all deposited at the bottom of some ancient sea.

The date of their formation no man yet can tell, but that it was vastly distant is certain. For the geological era is not over. Aqueous action still goes on: still does frost chip the rocks into fragments; still do mountain torrents sweep stone and mud and *débris* down the gulleys and watercourses; still do rivers erode their channels, and carry mud and silt far out to sea. And, more powerful than any of these agents of denudation, the waves and the tides are still at work along every coast-line, eating away into the cliffs, undermining gradually and submerging acre after acre, and making with the refuse a shingly, or a sandy, or a muddy beach—the nucleus of a new geological formation.

Of all denuding agents, there can be no doubt that, to the land exposed to them, the waves of the sea are by far the most powerful. Think how they beat and tear, and drive and drag, until even the hardest rock, like basalt, becomes honeycombed into strange galleries and passages—Fingal's Cave, for instance—and the softer parts are crumbled away. But the area now exposed to the teeth of the waves is not great. The fury of a winter storm may dash them a little higher than usual, but they cannot reach cliffs 100 feet high. They can undermine such cliffs indeed, and then grind the fragments to powder, but their direct action is limited. Not so limited, however, as they would be without the tides. Consider for a moment the denudation import of the tides: how does the existence of tidal rise and fall affect the geological problem?

The scouring action of the tidal currents themselves is not to be despised. It is the tidal ebb and flow which keeps open channel in the Mersey, for instance. But few places are so favourably situated as Liverpool in this respect, and the direct scouring action of the tides in general is not very great. Their geological import mainly consists in this—that they raise and lower the surface waves at regular intervals, so as to apply them to a considerable stretch of coast. The waves are a great planing machine attacking the land, and the tides raise and lower this planing machine, so that its denuding tooth is applied, now twenty feet vertically above mean level, now twenty feet below.

Making all allowance for the power of winds and waves, currents, tides, and watercourses, assisted by glacial ice and frost, it must be apparent how slowly the work of forming the rocks is being carried on. It goes on steadily, but so slowly that it is estimated to take 6000 years to wear away one foot of the American continent by all the denuding causes combined. To erode a stratum 5000 feet thick will require at this rate thirty million years.

The age of the earth is not at all accurately known, but there are many grounds for believing it not to be much older than some thirty million years. That is to say, not greatly more than this period of time has elapsed since it was in a molten condition. It may be as old as a hundred million years, but its age is believed by those most competent to judge to be more likely within this limit than beyond it. But if we ask what is the thickness of the rocks which in past times have been formed, and denuded, and re-formed, over and over again, we get an answer, not in feet, but in miles. The Laurentian and Huronian rocks of Canada constitute a stratum ten miles thick; and everywhere the rocks at the base of our stratified system are of the most stupendous volume and thickness.

It has always been a puzzle how known agents could have formed these mighty masses, and the only solution offered by geologists was, unlimited time. Given unlimited time, they could, of course, be formed, no matter how slowly the process went on. But inasmuch as the time allowable since the earth was cool enough for water to exist on it except as steam is not by any means unlimited, it becomes necessary to look for a far more powerful engine than any now existing; there must have been some denuding agent in those remote ages—ages far more distant from us than the Carboniferous period, far older than any forms of life, fossil or otherwise, ages among the oldest known to geology—a denuding agent must have then existed, far more powerful than any we now know.

Such an agent it has been the privilege of astronomy and physics, within the last ten years, to discover. To this discovery I now proceed to lead up.

Our fundamental standard of time is the period of the earth's rotation—the length of the day. The earth is our one standard clock: all time is expressed in terms of it, and if it began to go wrong, or if it did not go with perfect uniformity, it would seem a most difficult thing to discover its error, and a most puzzling piece of knowledge to utilize when found.

That it does not go much wrong is proved by the fact that we can calculate back to past astronomical events—ancient eclipses and the like—and we find that the record of their occurrence, as made by the old magi of Chaldæa, is in very close accordance with the result of calculation. One of these famous old eclipses was observed in

Babylon about thirty-six centuries ago, and the Chaldæan astronomers have put on record the time of its occurrence. Modern astronomers have calculated back when it should have occurred, and the observed time agrees very closely with the actual, but not exactly. Why not exactly?

Partly because of the acceleration of the moon's mean motion, as explained in the lecture on Laplace (p. 262). The orbit of the earth was at that time getting rounder, and so, as a secondary result, the speed of the moon was slightly increasing. It is of the nature of a perturbation, and is therefore a periodic not a progressive or continuous change, and in a sufficiently long time it will be reversed. Still, for the last few thousand years the moon's motion has been, on the whole, accelerated (though there seems to be a very slight retarding force in action too).

Laplace thought that this fact accounted for the whole of the discrepancy; but recently, in 1853, Professor Adams re-examined the matter, and made a correction in the details of the theory which diminishes its effect by about one-half, leaving the other half to be accounted for in some other way. His calculations have been confirmed by Professor Cayley. This residual discrepancy, when every known cause has been allowed for, amounts to about one hour.

The eclipse occurred later than calculation warrants. Now this would have happened from either of two causes, either an acceleration of the moon in her orbit, or a retardation of the earth in her diurnal rotation—a shortening of the month or a lengthening of the day, or both. The total discrepancy being, say, two hours, an acceleration of six seconds-per-century per century will in thirty-six centuries amount to one hour; and this, according to the corrected Laplacian theory, is what has occurred. But to account for the other hour some other cause must be sought, and at present it is considered most probably due to a steady retardation of the earth's rotation—a slow, very slow, lengthening of the day.

The statement that a solar eclipse thirty-six centuries ago was an hour late, means that a place on the earth's surface came into the shadow one hour behind time—that is, had lagged one twenty-fourth part of a revolution. The earth, therefore, had lost this amount in the course of $3600 \times 365\frac{1}{4}$ revolutions. The loss per revolution is exceedingly small, but it accumulates, and at any era the total loss is the sum of all the losses preceding it. It may be worth while just to explain this point further.

Suppose the earth loses a small piece of time, which I will call an instant, per day; a locality on the earth will come up to a given position one instant late on the first day after an event. On the next day it would come up two instants late by reason of the previous loss; but it also loses another instant during the course of the second day, and so the total lateness by the end of that day amounts to three instants. The day after, it will be going slower from the beginning at the rate of two instants a day, it will lose another instant on the fresh day's own account, and it started three instants late; hence the aggregate loss by the end of the third day is $1 + 2 + 3 = 6$. By the end of the fourth day the whole loss will be $1 + 2 + 3 + 4$, and so on. Wherefore by merely losing one instant every day the total loss in n days is $(1 + 2 + 3 + ... + n)$ instants, which amounts to $\frac{1}{2}n(n + 1)$ instants; or practically, when n is big, to $\frac{1}{2}n^2$. Now in thirty-six centuries there have been $3600 \times 365\frac{1}{4}$ days, and the total loss has amounted to an hour; hence the length of "an instant," the loss per diem, can be found from the equation $\frac{1}{2}(3600 \times 365)^2$ instants = 1 hour; whence one "instant" equals the 240 millionth part of a second. This minute quantity represents the retardation of the earth per day. In a year the aggregate loss mounts up to $\frac{1}{3600}$th part of a second, in a century to about three seconds, and in thirty-six centuries to an hour. But even at the end of the thirty-six centuries the day is barely any longer; it is only 3600 × 365 instants, that is 1/180th of a second, longer than it was at the beginning. And even a million years ago, unless the rate of loss was different (as it probably was), the day would only be thirty-five minutes shorter, though by that time the aggregate loss, as measured by the apparent lateness of any perfectly punctual event reckoned now, would have amounted to nine years. (These numbers are to be taken as illustrative, not as precisely representing terrestrial fact.)

What can have caused the slowing down? Swelling of the earth by reason of accumulation of meteoric dust might do something, but probably very little. Contraction of the earth as it goes on cooling would act in the opposite direction, and probably more than counterbalance the dust effect. The problem is thus not a simple one, for there are several disturbing causes, and for none of them are the data enough to base a quantitative estimate upon; but one certain agent in lengthening the day, and almost certainly the main agent, is to be found in the tides.

Remember that the tidal humps were produced as the prolateness of a sphere whirled round and round a fixed centre, like a football whirled by a string. These humps are pulled at by the moon, and the earth rotates on its axis against this pull. Hence it tends to be constantly, though very slightly, dragged back.

In so far as the tidal wave is allowed to oscillate freely, it will swing with barely any maintaining force, giving back at one quarter-swing what it has received at the previous quarter; but in so far as it encounters friction, which it does in all channels where there is an actual ebb and flow of the water, it has to receive more than it gives back, and the balance of energy has to be made up to it, or the tides would cease. The energy of the tides is, in fact, continually being dissipated by friction, and all the energy so dissipated is taken from the rotation of the earth. If tidal energy were utilized by engineers, the machines driven would be really driven at the expense of the earth's rotation: it would be a mode of harnessing the earth and using the moon as fixed point or fulcrum; the moon pulling at the tidal protuberance, and holding it still as the earth rotates, is the mechanism whereby the energy is extracted, the handle whereby the friction brake is applied.

Winds and ocean currents have no such effect (as Mr. Fronde in *Oceania* supposes they have), because they are all accompanied by a precisely equal counter-current somewhere else, and no internal rearrangement of fluid can affect the motion of a mass as a whole; but the tides are in different case, being produced, not by internal inequalities of temperature, but by a straightforward pull from an external body.

The ultimate effect of tidal friction and dissipation of energy will, therefore, be to gradually retard the earth till it does not rotate with reference to the moon, *i.e.* till it rotates once while the moon revolves once; in other words, to make the day and the month equal. The same cause must have been in operation, but with eighty-fold greater intensity, on the moon. It has ceased now, because the rotation has stopped, but if ever the moon rotated on its axis with respect to the earth, and if it were either fluid itself or possessed any liquid ocean, then the tides caused by the pull of the earth must have been prodigious, and would tend to stop its rotation. Have they not succeeded? Is it not probable that this is *why* the moon always now turns the same face towards us? It is believed to be almost certainly the cause. If so, there was a time when the moon behaved differently—when it rotated more quickly than it revolved, and exhibited to us its whole surface. And at this era, too, the earth itself must have rotated a little faster, for it has been losing speed ever since.

We have thus arrived at this fact, that a thousand years ago the day was a trifle shorter than it is now. A million years ago it was, perhaps, an hour shorter. Twenty million years ago it must have been much shorter. Fifty million years ago it may have been only a few hours long. The earth may have spun round then quite quickly. But there is a limit. If it spun too fast it would fly to pieces. Attach shot by means of wax to the whirling earth model, Fig. 110, and at a certain speed the cohesion of the wax cannot hold them, so they fly off. The earth is held together not by cohesion but by gravitation; it is not difficult to reckon how fast the earth must spin for gravity at its surface to be annulled, and for portions to fly off. We find it about one revolution in three hours. This is a critical speed. If ever the day was three hours long, something must have happened. The day can never have been shorter than that; for if it were, the earth would have a tendency to fly in pieces, or, at least, to separate into two pieces. Remember this, as a natural result of a three-hour day, which corresponds to an unstable state of things; remember also that in some past epoch a three-hour day is a probability.

If we think of the state of things going on in the earth's atmosphere, if it had an atmosphere at that remote date, we shall recognize the existence of the most fearful tornadoes. The trade winds, which are now peaceful agents of commerce, would then be perpetual hurricanes, and all the denudation agents of the geologist would be in a state of feverish activity.

So, too, would the tides: instead of waiting six hours between low and high tide, we should have to wait only three-quarters of an hour. Every hour-and-a-half the water would execute a complete swing from high tide to high again.

Very well, now leave the earth, and think what has been happening to the moon all this while.

We have seen that the moon pulls the tidal hump nearest to it back; but action and reaction are always equal and opposite—it cannot do that without itself getting pulled forward. The pull of the earth on the moon will therefore not be quite central, but will be a little in advance of its centre; hence, by Kepler's second law, the rate of description of areas by its radius vector cannot be constant, but must increase (p. 208). And the way it increases will be for the radius vector to lengthen, so as to sweep out a bigger area. Or, to put it another way, the extra speed tending to be gained by the moon will fling it further away by extra centrifugal force. This last is not so good a way of regarding the matter; though it serves well enough for the case of a ball whirled at the end of an elastic string. After having got up the whirl, the hand holding the string may remain almost fixed at the centre of the circle, and the motion will continue steadily; but if the hand be moved so as always to pull the string a little in advance of the centre, the speed of whirl will increase, the elastic will be more and more stretched, until the whirling ball is describing a much larger circle. But in this case it will likewise be going faster—distance and speed increase together. This is because it obeys a different law from gravitation—the force is not inversely as the square, or any other single power, of the distance. It does not obey any of Kepler's laws, and so it does not obey the one which now concerns us, viz. the third; which practically states that the further a planet is from the centre the slower it goes; its velocity varies inversely with the square root of its distance (p. 74).

If, instead of a ball held by elastic, it were a satellite held by gravity, an increase in distance must be accompanied by a diminution in speed. The time of revolution varies as the square of the cube root of the distance (Kepler's third law). Hence, the tidal reaction on the moon, having as its primary effect, as we have seen, the pulling the moon a little forward, has also the secondary or indirect effect of making it move slower and go further off. It may seem strange that an accelerating pull, directed in front of the centre, and therefore always pulling the moon the way it is going, should retard it; and that a retarding force like friction, if such a force acted, should hasten it, and make it complete its orbit sooner; but so it precisely is.

Gradually, but very slowly, the moon is receding from us, and the month is becoming longer. The tides of the earth are pushing it away. This is not a periodic disturbance, like the temporary acceleration of its motion discovered by Laplace, which in a few centuries, more or less, will be reversed; it is a disturbance which always acts one way, and which is therefore cumulative. It is superposed upon all periodic changes, and, though it seems smaller than they, it is more inexorable. In a thousand years it makes scarcely an appreciable change, but in a million years its persistence tells very distinctly; and so, in the long run, the month is getting longer and the moon further off. Working backwards also, we see that in past ages the moon must have been nearer to us than it is now, and the month shorter.

Now just note what the effect of the increased nearness of the moon was upon our tides. Remember that the tide-generating force varies inversely as the cube of distance, wherefore a small change of distance will produce a great difference in the tide-force.

The moon's present distance is 240 thousand miles. At a time when it was only 190 thousand miles, the earth's tides would have been twice as high as they are now. The pushing away action was then a good deal more violent, and so the process went on quicker. The moon must at some time have been just half its present distance, and the tides would then have risen, not 20 or 30 feet, but 160 or 200 feet. A little further back still, we have the moon at one-third of its present distance from the earth, and the tides 600 feet high. Now just contemplate the effect of a 600-foot tide. We are here only about 150 feet above the level of the sea; hence, the tide would sweep right over us and rush far away inland. At high tide we should have some 200 feet of blue water over our heads. There would be nothing to stop such a tide as that in this neighbourhood till it reached the high lands of Derbyshire. Manchester would be a seaport then with a vengeance!

The day was shorter then, and so the interval between tide and tide was more like ten than twelve hours. Accordingly, in about five hours, all that mass of water would have swept back again, and great tracts of sand between here and Ireland would be left dry. Another five hours, and the water would come tearing and driving over the country, applying its furious waves and currents to the work of denudation, which would proceed apace. These high tides of enormously distant past ages constitute the denuding agent which the geologist required. They are very ancient—more ancient than the Carboniferous period, for instance, for no trees could stand the furious storms that must have been prevalent at this time. It is doubtful whether any but the very lowest forms of life then existed. It is the strata at the bottom of the geological scale that are of the most portentous thickness, and the only organism suspected in them is the doubtful *Eozoon Canadense.* Sir Robert Ball believes, and several geologists agree with him, that the mighty tides we are contemplating may have been coæval with this ancient Laurentian formation, and others of like nature with it.

But let us leave geology now, and trace the inverted progress of events as we recede in imagination back through the geological era, beyond, into the dim vista of the past, when the moon was still closer and closer to the earth, and was revolving round it quicker and quicker, before life or water existed on it, and when the rocks were still molten.

Suppose the moon once touched the earth's surface, it is easy to calculate, according to the principles of gravitation, and with a reasonable estimate of its size as then expanded by heat, how fast it must then have revolved round the earth, so as just to save itself from falling in. It must have gone round once every three hours. The month was only three hours long at this initial epoch.

Remember, however, the initial length of the day. We found that it was just possible for the earth to rotate on its axis in three hours, and that when it did so, something was liable to separate from it. Here we find the moon in contact with it, and going round it in this same three-hour period. Surely the two are connected. Surely the moon was a part of the earth, and was separating from it.

That is the great discovery—the origin of the moon.

Once, long ages back, at date unknown, but believed to be certainly as much as fifty million years ago, and quite possibly one hundred million, there was no moon, only the earth as a molten globe, rapidly spinning on its axis—spinning in about three hours. Gradually, by reason of some disturbing causes, a protuberance, a sort of bud, forms at one side, and the great inchoate mass separates into two—one about eighty times as big as the other. The bigger one we now call earth, the smaller we now call moon. Round and round the two bodies went, pulling each other into tremendously elongated or prolate shapes, and so they might have gone on for a long time. But they are unstable, and cannot go on thus: they must either separate or collapse. Some disturbing cause acts again, and the smaller mass begins to revolve less rapidly. Tides at once begin—gigantic tides of molten lava hundreds of miles high; tides not in free ocean, for there was none then, but in the pasty mass of the entire earth. Immediately the series of changes I have described begins, the speed of rotation gets slackened, the moon's mass gets pushed further and further away, and its time of revolution grows rapidly longer. The changes went on rapidly at first, because the tides were so gigantic; but gradually, and by slow degrees, the bodies get more distant, and the rate of change more moderate. Until, after the lapse of ages, we find the day twenty-four hours long, the moon 240,000 miles distant, revolving in 27⅓ days, and the tides only existing in the water of the ocean, and only a few feet high. This is the era we call "to-day."

The process does not stop here: still the stately march of events goes on; and the eye of Science strives to penetrate into the events of the future with the same clearness as it has been able to descry the events of the past. And what does it see? It will take too long to go into full detail: but I will shortly summarize the results. It sees this first—the day and the month both again equal, but both now about 1,400 hours long. Neither of these bodies rotating with respect to each other—the two as if joined by a bar—and total cessation of tide-generating action between them.

The date of this period is one hundred and fifty millions of years hence, but unless some unforeseen catastrophe intervenes, it must assuredly come. Yet neither will even this be the final stage; for the system is disturbed by the tide-generating force of the sun. It is a small effect, but it is cumulative; and gradually, by much slower degrees than anything we have yet contemplated, we are presented with a picture of the month getting gradually shorter than the day, the moon gradually approaching instead of receding, and so, incalculable myriads of ages hence, precipitating itself upon the surface of the earth whence it arose.

Such a catastrophe is already imminent in a neighbouring planet—Mars. Mars' principal moon circulates round him at an absurd pace, completing a revolution in 7½ hours, and it is now only 4,000 miles from his surface. The planet rotates in twenty-four hours as we do; but its tides are following its moon more quickly than it rotates after them; they are therefore tending to increase its rate of spin, and to retard the revolution of the moon. Mars is therefore slowly but surely pulling its moon down on to itself, by a reverse action to that which separated our moon. The day shorter than the month forces a moon further away; the month shorter than the day tends to draw a satellite nearer.

This moon of Mars is not a large body: it is only twenty or thirty miles in diameter, but it weighs some forty billion tons, and will ultimately crash along the surface with a velocity of 8,000 miles an hour. Such a blow must produce the most astounding effects when it occurs, but I am unable to tell you its probable date.

So far we have dealt mainly with the earth and its moon; but is the existence of tides limited to these bodies? By no means. No body in the solar system is rigid, no body in the stellar universe is rigid. All must be susceptible of some tidal deformation, and hence, in all of them, agents like those we have traced in the history of the earth and moon must be at work: the motion of all must be complicated by the phenomena of tides. It is Prof. George Darwin who has worked out the astronomical influence of the tides, on the principles of Sir William Thomson: it is Sir Robert Ball who has extended Mr. Darwin's results to the past history of our own and other worlds.[32]

Tides are of course produced in the sun by the action of the planets, for the sun rotates in twenty-five days or thereabouts, while the planets revolve in much longer periods than that. The principal tide-generating bodies will be Venus and Jupiter; the greater nearness of one rather more than compensating for the greater mass of the other.

It may be interesting to tabulate the relative tide-producing powers of the planets on the sun. They are as follows, calling that of the earth 1,000:—

Relative Tide-producing Powers of the Planets on the Sun.

Mercury	1,121
Venus	2,339
Earth	1,000
Mars	304
Jupiter	2,136
Saturn	1,033
Uranus	21
Neptune	9

The power of all of them is very feeble, and by acting on different sides they usually partly neutralize each other's action; but occasionally they get all on one side, and in that case some perceptible effect may be produced; the probable effect seems likely to be a gentle heaving tide in the solar surface, with breaking up of any incipient crust; and such an effect may be considered as evidenced periodically by the great increase in the number of solar spots which then break out.

The solar tides are, however, much too small to appreciably push any planet away, hence we are not to suppose that the planets originated by budding from the sun, in contradiction of the nebular hypothesis. Nor is it necessary to assume that the satellites, as a class, originated in the way ours did; though they may have done so. They were more probably secondary rings. Our moon differs from other satellites in being exceptionally large compared with the size of its primary; it is as big as some of the moons of Jupiter and Saturn. The earth is the only one of the small planets that has an appreciable moon, and hence there is nothing forced or unnatural in supposing that it may have had an exceptional history.

Evidently, however, tidal phenomena must be taken into consideration in any treatment of the solar system through enormous length of time, and it will probably play a large part in determining its future.

When Laplace and Lagrange investigated the question of the stability or instability of the solar system, they did so on the hypothesis that the bodies composing it were rigid. They reached a grand conclusion—that all the mutual perturbations of the solar system were periodic—that whatever changes were going on would reach a maximum and then begin to diminish; then increase again, then diminish, and so on. The system was stable, and its changes were merely like those of a swinging pendulum.

But this conclusion is not final. The hypothesis that the bodies are rigid is not strictly true: and directly tidal deformation is taken into consideration it is perceived to be a potent factor, able in the long run to upset all their calculations. But it is so utterly and inconceivably minute—it only produces an appreciable effect after millions of years—whereas the ordinary perturbations go through their swings in some hundred thousand years or so at the most. Granted it is small, but it is terribly persistent; and it always acts in one direction. Never does it cease: never does it begin to act oppositely and undo what it has done. It is like the perpetual dropping of water. There may be only one drop in a twelvemonth, but leave it long enough, and the hardest stone must be worn away at last.

We have been speaking of millions of years somewhat familiarly; but what, after all, is a million years that we should not speak familiarly of it? It is longer than our lifetime, it is true. To the ephemeral insects whose lifetime is an hour, a year might seem an awful period, the mid-day sun might seem an almost stationary body, the changes of the seasons would be unknown, everything but the most fleeting and rapid changes would appear permanent and at rest. Conversely, if our life-period embraced myriads of æons, things which now seem permanent would then appear as in a perpetual state of flux. A continent would be sometimes dry, sometimes covered with ocean; the stars we now call fixed would be moving visibly before our eyes; the earth would be humming on its axis like a top, and the whole of human history might seem as fleeting as a cloud of breath on a mirror.

Evolution is always a slow process. To evolve such an animal as a greyhound from its remote ancestors, according to Mr. Darwin, needs immense tracts of time; and if the evolution of some feeble animal crawling on the surface of this planet is slow, shall the stately evolution of the planetary orbs themselves be hurried? It may be that we are able to trace the history of the solar system for some thousand million years or so; but for how much longer time must it not have a history—a history, and also a future—entirely beyond our ken?

Those who study the stars have impressed upon them the existence of the most immeasurable distances, which yet are swallowed up as nothing in the infinitude of space. No less are we compelled to recognize the existence of incalculable æons of time, and yet to perceive that these are but as drops in the ocean of eternity.

FOOTNOTES:

[1] The following account of Mars's motion is from the excellent small manual of astronomy by Dr. Haughton of Trinity College, Dublin:—(P. 151) "Mars's motion is very unequal; when he first appears in the morning emerging from the rays of the sun, his motion is direct and rapid; it afterwards becomes slower, and he becomes stationary when at an elongation of 137° from the sun; then his motion becomes retrograde, and its velocity increases until he is in opposition to the sun at 180°; at this time the retrograde motion is most rapid, and afterwards diminishes until he is 137° distant from the sun on the other side, when Mars again becomes stationary; his motion then becomes direct, and increases in velocity until it reaches a maximum, when the planet is again in conjunction with the sun. The retrograde motion of this planet lasts for 73 days: and its arc of retrogradation is 16°."

[2] It is not so easy to plot the path of the sun among the stars by direct observation, as it is to plot the path of a planet; because sun and stars are not visible together. Hipparchus used the moon as an intermediary; since sun and moon are visible together, and also moon and stars.

[3] This is, however, by no means the whole of the matter. The motion is not a simple circle nor has it a readily specifiable period. There are several disturbing causes. All that is given here is a first rough approximation.

[4] The proof is easy, and ought to occur in books on solid geometry. By a "regular" solid is meant one with all its faces, edges, angles, &c., absolutely alike: it is of these perfectly symmetrical bodies that there are only five. Crystalline forms are practically infinite in number.

[5] Best known to us by his Christian name, as so many others of that time are known, *e.g.* Raphael Sanzio, Dante Alighieri, Michael Angelo Buonarotti. The rule is not universal. Tasso and Ariosto are surnames.

[6] It would seem that the fact that all bodies of every material tend to fall at the same rate is still not clearly known. Confusion is introduced by the resistance of the air. But a little thought should make it clear that the effect of the air is a mere disturbance, to be eliminated as far as possible, since the atmosphere has nothing to do with gravitation. The old fashioned "guinea and feather experiment" illustrates that in a vacuum things entirely different in specific gravity or surface drop at the same pace.

[7] Karl von Gebler (Galileo), p. 13.

[8] It is of course the "silver lining" of clouds that outside observers see.

[9] L.U.K., *Life of Galileo*, p. 26.

[10] *Note added September, 1892.* News from the Lick Observatory makes a very small fifth satellite not improbable.

[11] They remained there till this century. In 1835 they were quietly dropped.

[12] It was invented by van Helmont, a Belgian chemist, who died in 1644. He suggested two names *gas* and *blas*, and the first has survived. Blas was, I suppose, from *blasen*, to blow, and gas seems to be an attempt to get at the Sanskrit root underlying all such words as *geist*.

[13] Such as this, among many others:—The duration of a flame under different conditions is well worth determining. A spoonful of warm spirits of wine burnt 116 pulsations. The same spoonful of spirits of wine with addition of one-sixth saltpetre burnt 94 pulsations. With one-sixth common salt, 83; with one-sixth gunpowder, 110; a piece of wax in the

middle of the spirit, 87; a piece of *Kieselstein*, 94; one-sixth water, 86; and with equal parts water, only 4 pulse-beats. This, says Liebig, is given as an example of a "*licht-bringende Versuch*."

[14] Draper, *History of Civilization in Europe*, vol. ii. p. 259.

[15] Professor Knight's series of Philosophical Classics.

[16] To explain why the entire system, horse and cart together, move forward, the forces acting on the ground must be attended to.

[17] The distance being proportional to the *square* of the time, see p. 82.

[18] The following letter, recently unearthed and published in *Nature*, May 12, 1881, seems to me well worth preserving. The feeling of a respiratory interval which it describes is familiar to students during the too few periods of really satisfactory occupation. The early guess concerning atmospheric electricity is typical of his extraordinary instinct for guessing right.

"LONDON, *Dec. 15, 1716*.

"DEAR DOCTOR,—He that in ye mine of knowledge deepest diggeth, hath, like every other miner, ye least breathing time, and must sometimes at least come to terr. alt. for air.

"In one of these respiratory intervals I now sit down to write to you, my friend.

"You ask me how, with so much study, I manage to retene my health. Ah, my dear doctor, you have a better opinion of your lazy friend than he hath of himself. Morpheous is my last companion; without 8 or 9 hours of him yr correspondent is not worth one scavenger's peruke. My practices did at ye first hurt my stomach, but now I eat heartily enou' as y' will see when I come down beside you.

"I have been much amused at ye singular φενόμενα resulting from bringing of a needle into contact with a piece of amber or resin fricated on silke clothe. Ye flame putteth me in mind of sheet lightning on a small—how very small—scale. But I shall in my epistles abjure Philosophy whereof when I come down to Sakly I'll give you enou'. I began to scrawl at 5 mins. from 9 of ye clk. and have in writing consmd. 10 mins. My Ld. Somerset is announced.

"Farewell, Gd. bless you and help yr sincere friend.

"ISAAC NEWTON.

"*To* DR. LAW, Suffolk."

[19] Kepler's laws may be called respectively, the law of path, the law of speed, and the relationship law. By the "mass" of a body is meant the number of pounds or tons in it: the amount of matter it contains. The idea is involved in the popular word "massive."

[20] The equation we have to verify is

$$gR^2 = \frac{4\pi^2 r^3}{T^2},$$

with the data that r, the moon's distance, is 60 times R, the earth's radius, which is 3,963 miles; while T, the time taken to complete the moon's orbit, is 27 days, 13 hours, 18 minutes, 37 seconds. Hence, suppose we calculate out g, the intensity of terrestrial gravity, from the above equation, we get

$$g = \frac{4\pi^2}{T^2} \times (60)^3 = \frac{39{\cdot}92 \times 216000 \times 3963 \text{ miles}}{(27 \text{ days, } 13 \text{ hours, \&c.})^2} = 32{\cdot}92 \text{ feet-per-second per second,}$$

which is not far wrong.

[21] The two motions may be roughly compounded into a single motion, which for a few centuries may without much error be regarded as a conical revolution about a different axis with a different period; and Lieutenant-Colonel Drayson writes books emphasizing this simple fact, under the impression that it is a discovery.

[22] Members of the Accademia dei Lyncei, the famous old scientific Society established in the time of Cosmo de Medici—older than our own Royal Society.

[23] Newton suspected that the moon really did so oscillate, and so it may have done once; but any real or physical libration, if existing at all, is now extremely minute.

[24] An interesting picture in the New Gallery this year (1891), attempting to depict "Earth-rise in Moon-land," unfortunately errs in several particulars. First of all, the earth does not "rise," but is fixed relatively to each place on the moon; and two-fifths of the moon never sees it. Next, the earth would not look like a map of the world with a haze on its edge. Lastly, whatever animal remains the moon may contain would probably be rather in the form of fossils than of skeletons. The skeleton is of course intended as an image of death and desolation. It is a matter of taste: but a skeleton, it seems to me, speaks too recently of life to be as appallingly weird and desolate as a blank stone or ice landscape, unshaded by atmosphere or by any trace of animal or plant life, could be made.

[25] Five of Jupiter's revolutions occupy 21,663 days; two of Saturn's revolutions occupy 21,526 days.

[26] *Excircularity* is what is meant by this term. It is called "excentricity" because the foci (not the centre) of an ellipse are regarded as the representatives of the centre of a circle. Their distance from the centre, compared with the radius of the unflattened circle, is called the excentricity.

[27] A curve of the *n*th degree has ½$n(n+3)$ arbitrary constants in its equation, hence this number of points specifically determine it. But special points, like focus or vertex, count as two ordinary ones. Hence three points plus the focus act as five points, and determine a conic or curve of the second degree. Three observations therefore fix an orbit round the sun.

[28] Its name suggests a measure of the diameter of the sun's disk, and this is one of its functions; but it can likewise measure planetary and other disks; and in general behaves as the most elaborate and expensive form of micrometer. The Königsberg instrument is shewn in fig. 92.

[29] It may be supposed that the terms "minute" and "second" have some necessary connection with time, but they are mere abbreviations for *partes minutæ* and *partes minutæ secundæ*, and consequently may be applied to the subdivision of degrees just as properly as to the subdivision of hours. A "second" of arc means the 3600th part of a degree, just as a second of time means the 3600th part of an hour.

[30] A group of flying particles, each one invisible, obstructs light singularly little, even when they are close together, as one can tell by the transparency of showers and snowstorms. The opacity of haze may be due not merely to dust particles, but to little eddies set up by radiation above each particle, so that the air becomes turbulent and of varying density. (See a similar suggestion by Mr. Poynting in *Nature*, vol. 39, p. 323.)

[31] The moon ought to be watched during the next great shower, if the line of fire happens to take effect on a visible part of the dark portion.

[32] Address to Birmingham Midland Institute, "A Glimpse through the Corridors of Time."